Agrarian Transformations

Agrarian Transformations

Local Processes and the State in Southeast Asia

EDITED BY
GILLIAN HART, ANDREW TURTON,
AND BENJAMIN WHITE
WITH BRIAN FEGAN
AND LIM TECK GHEE

UNIVERSITY OF CALIFORNIA PRESS
BERKELEY LOS ANGELES LONDON

University of California Press
Berkeley and Los Angeles, California

University of California Press, Ltd.
London, England

Library of Congress Cataloging-in-Publication Data

Agrarian transformations : local processes and the state in Southeast Asia / edited by Gillian Hart, Andrew Turton, and Benjamin White, with Brian Fegan and Lim Teck Ghee.
p. cm.
Includes index.
ISBN 0-520-06197-7 (alk. paper)
1. Agriculture and state—Asia, Southeastern—Case studies.
2. Land tenure—Asia, Southeastern—Case studies. 3. Agricultural innovations—Economic aspects—Asia, Southeastern—Case studies.
I. Hart, Gillian Patricia. II. Turton, Andrew. III. White, Benjamin.
HD2075.5.Z8A37 1989
338.1'0959—dc19 88-10626
CIP

Printed in the United States of America
1 2 3 4 5 6 7 8 9

Contents

Maps

Tables

Preface

This book is part of an effort to synthesize and stimulate research on agrarian change in Southeast Asia that was initiated and supported by the Joint Committee on Southeast Asia (JCSEA) of the Social Science Research Council (SSRC) and the American Council of Learned Societies with funds from the Ford Foundation. Its purpose was twofold: to integrate and compare evidence on the rapid and profound agrarian changes taking place in different countries in the region, and to address the conceptual and methodological problems involved in understanding these changes.

With these broad purposes in mind, the JCSEA convened a planning meeting at the University of Washington in Seattle in 1981 to define a set of topics around which more specific synthetic efforts could be constructed. Four such topics emerged: Everyday Forms of Peasant Resistance; Consumption, Communication, and Culture; National Political-Economic Strategies; and the present study, initially defined as "Issues and Processes of Rural Differentiation."

As originally conceived, the "rural differentiation" project was to identify patterns and processes of change in the control of resources and power in contemporary rural Southeast Asia and to relate these patterns and processes to theories of agrarian change and differentiation. We were concerned with both the general tendencies emerging in different areas as rural producers became more tightly enmeshed in larger circuits of exchange and the particular forms of structural change in different areas.

Both for the sake of comparison and for reasons discussed more fully in the introduction, we decided to draw upon detailed local-level studies of the main rice-growing regions of Indonesia, Malaysia, the Philippines, Thailand, and Vietnam. As the project evolved, we became increasingly concerned with the conceptual and methodological problems of linking local-level institutional arrangements with larger political-economic systems. Although we started out with a strong sense of the importance of these connections, in practice most of the early drafts of the papers dealt almost exclusively with the local level. Accordingly, questions of how to link the local-level studies to larger systems assumed growing importance.

In time, the exercise of power at different levels of society emerged as a central element of our efforts to understand these connections and to account for patterns of similarity and variation. Far from a preconceived notion applied to different cases, the emphasis on the exercise of power in different historical contexts developed in the lengthy process of our discussing and rewriting the papers, elaborating those studies where more qualitative evidence was available, and constructing introductions to the case studies for each country.

Focusing on the exercise of power not only provided a key to relating the material contained in the local-level studies to larger political-economic forces: it has also served to highlight gaps and suggest future directions for agrarian research. One of the most important of these concerns was the need to examine relations between genders and generations as both a cause and a consequence of the exercise of power and the dynamics of social change.

In this book we have tried to achieve as much coherence as possible, but it must be emphasized that the case studies included here represent work that was already completed or underway at the time this project was conceived. Our efforts to attain coherence required considerable sacrifice in terms of studies that had to be excluded. Our original plan included two studies of Vietnam with the other four countries, and Ngo Vinh Long and Adam Fforde contributed extremely interesting papers on agrarian conditions in southern and northern Vietnam respectively. Unfortunately, the type of evidence available on Vietnam is sufficiently different from the local-level studies around which the central themes of this book have developed that we had to exclude the Vietnamese case. The focus on rice-growing communities meant that we also had to exclude an excellent study by Willem Wolters of an upland village in Java.

One extremely useful aspect of our project was the participation of those scholars with similar thematic interests but whose own empirical work has focused on other world areas. The stimulating and constructive comments and suggestions of Utsa Patnaik, Alain de Janvry, and Sara Berry (respectively from a South Asian, Latin American, and African perspective), together with those of two anonymous reviewers of the University of California Press, are reflected in various parts of the book and are gratefully acknowledged. We are also grateful to Chai-anan Samudavanija, who made arrangements for a project workshop in Thailand.

As will be clear from this description of the process by which the "rural differentiation" project sought and eventually found its shape, the emergence and completion of the present volume would not have been possible without a major commitment from the JCSEA, which supported the two international workshops at the East-West Center in Honolulu and the Northern Region Agricultural Training Center in Chiengmai and the editors' meeting in Aix-en-Provence (three pleasant locations which, hard though it may be to believe, were chosen to minimize the total travel cost for participants coming from various continents), as well as support for typing and other costs of the editing process. This generous support is gratefully acknowledged.

In particular we would like to express our special and sincere thanks to David Szanton, the SSRC's staff associate for the Joint Committee on Southeast Asia from 1976 to 1986. Without his encouragement, his wise judgment, and his friendly but firm admonitions, we doubt that our project would have come to term. This book, together with other Southeast Asian products of SSRC sponsorship, reflects the important role which David has played in Southeast Asian studies.

Introduction

In this volume we have brought together a set of studies with the purpose of better understanding local-level processes of agrarian transformation and differentiation in relation to wider political-economic systems. Although all the studies are drawn from major rice-growing regions in Indonesia, Malaysia, the Philippines, and Thailand, this is not primarily a book about rice in Southeast Asia. Rather, our chief concerns are twofold: to specify the mechanisms and processes of agrarian change and their linkages with larger political and economic forces, and to suggest how detailed local-level studies can contribute to a better conceptual understanding of the causes and consequences of agrarian change.

Wet rice is a crop of particular interest in the analysis of agrarian differentiation partly because of its alleged technological properties. The capacity of wet-rice agriculture to absorb huge amounts of labor at more or less constant levels of productivity is central to concepts of agricultural involution and shared poverty (Geertz 1963). More recently, Bray (1983, 1986) has maintained that diseconomies of scale in Asian wet-rice cultivation have precluded mechanization and the concentration of landholdings, thereby resulting in patterns of differentiation distinctively different from those in Europe. The evidence from rice-growing regions of Southeast Asia brought together in this book calls into question the notion that differentiation can be explained in terms of the technological peculiarities of wet rice.

The rapid spread of new seed and fertilizer technologies in Southeast Asia over the past fifteen years, along with the greater commercialization that they engender, constitute further reasons for focusing on rice-growing areas. Some writers in the neopopulist tradition claim that the Green Revolution is primarily responsible for growing inequality (e.g., Palmer 1977), while neoclassicists cite inadequate technological progress as the major cause (e.g., Hayami and Kikuchi 1982). In this book we stress the limitations of viewing rural change primarily in terms of technology and commercialization and emphasize the need to take explicit account of the power structures within which technological change and commercialization occur.

Efforts to conceptualize the dynamics of agrarian change and differentiation are more fully developed and actively debated in the Marxist and neo-Marxist literature, but many of these debates have been limited by their shaky empirical foundations. In some instances theorists have sought to defend a particular model by invoking indicators derived from conventional survey data that are often subject to alternative interpretations. In others, empirical referents have been either ignored or invoked in a highly stylized form. Thus, for example, dependency theorists and structuralists have generally been far more concerned with what is and isn't capitalist (and/or functional to it) than with understanding the dynamic processes at work in a particular setting.

In recent years there has been growing dissatisfaction with these debates, much of it focused on the functionalism and teleology that underpin them. From these and other criticisms, a consensus is beginning to emerge around the need to understand particular forms of production in relation to the larger systems of which they are part (e.g., Bernstein 1979; Friedmann 1980; Kahn 1981, 1982; Lehmann 1982; Chevalier 1982; Smith 1984). "Consensus" is perhaps too strong a term in light of some significant variations among these analyses. Although a number of important conceptual and methodological questions remain open, this broad stream of thinking does represent a constructive departure from the earlier debates.

Some of the contributors to this book take issue with certain of the authors just cited, and readers will see that they certainly do not all conform to a specific model. All agree, however, on the need for a conceptual framework in which the institutional arrangements governing access to and control over resources and people are linked with larger economic and political forces. All agree, too, that it is through detailed local-level studies that better conceptual understandings along these lines will develop, and this book represents our collective effort to move in this direction.

The book is divided into five parts. Part I contains essays by White and Hart discussing a range of conceptual and methodological questions both in general terms and with reference to the comparative Southeast Asian material contained in the rest of the book. The organization of these case studies reflects our concern with relating local-level processes to larger political-economic systems. Each set of case studies is preceded by a country introduction that outlines the distinctive features of national political and economic structures, focusing particularly on the diverse ways in which the state has intervened in the rice sector.

Pointing to the rigidity that characterizes many theoretical formulations of the "agrarian question" and rural differentiation, White (Chapter 1) stresses the need for flexibility in the analysis of concrete situations. There is no universal form or definition of rural differentiation whose dynamics or "laws of motion" can be grasped through abstract formulation. Nor is it productive, he argues, to counterpose empirical evidence against an often reductionist version of the classical models of differentiation and thereby conclude that Lenin, Chayanov, etc., "do not work" in such and such setting.

In the analysis of concrete situations, White suggests that a distinction be drawn between the process of differentiation itself and various aspects of that process—namely the causes, mechanisms, and indicators. Arguments about differentiation processes (or their absence) are typically based on inferences drawn from a set of quantitative indicators, with these inferences informed by preexisting frameworks or theories. Direct documentation of mechanisms of differentiation—which many of the local-level studies in this volume provide—is far more unusual and constitutes the basis for different interpretations of processes of change.

Most analyses of the causes of rural transformation and differentiation have tended to focus on commercialization and technology, regarding the state and questions of power as largely external to agrarian processes. Drawing mainly on the comparative material in the case studies, Hart (Chapter 2) stresses the need to take explicit account of state imperatives and the exercise of power at different levels of society. By focusing on state patronage of dominant rural groups in Southeast Asia within a comparative framework, she shows how actions by the state not only influence local-level mechanisms of labor control and accumulation but are also affected by them. Accordingly, the unintended consequences of state policy constitute important sources of change within rural society as well as in the system as a whole. Hart's essay suggests the importance of viewing rural change as a dialectical rather than a linear process.

The emphasis on the state and the exercise of power at different levels of society reappears in Turton's essay on Thailand (Chapter 4), which addresses some of the analytic issues and problems entailed in incorporating the state into an understanding of agrarian change and differentiation. Turton reminds us that the state in Thailand (as in the other three countries) is not a unified, monolithic entity pursuing a coherent set of policies. Nor does it exist primarily to serve the interests of the bourgeoisie. Instead it is a heterogeneous power bloc comprised of various class and nonclass elements who struggle among themselves for control and influence.

Although the Thai, Indonesian, Malaysian, and Philippine states are all characterized by a degree of "relative autonomy," the particular political and economic forms this autonomy takes differ in important ways, as does the power of the accumulating classes vis-à-vis the state apparatus. Indeed, the value of comparative study of the four Southeast Asian countries lies partly in important variations in the nature of the state and the particular forms of capitalist development in the four countries.

At the same time, all four states share an intense strategic concern with ensuring plentiful and cheap rice supplies, and all have intervened actively in the rice sector, particularly since the widespread availability of modern rice technology from the late 1960's. However, the specific ways in which states have intervened to promote rice production and procurement have varied significantly. Malaysia and Indonesia have subsidized rice production, whereas Thailand has engaged in net taxation and, particularly since the late 1970's, the Philippines has used price controls and import policies to transfer wealth from rice producers to consumers and the urban classes.

On the face of it, these differential patterns of subsidy and taxation appear to reflect macroeconomic conditions and the fiscal capacity of different states in the region. Malaysia's relative wealth together with the small size of the rice sector have enabled the state to subsidize rice producers, whereas the Thai economy is still heavily dependent on extracting resources from agriculture and the state draws a large proportion of its revenues from rice. Furthermore, shifts in policy have often been far more closely associated with changing macroeconomic conditions than with changes in the political influence of dominant rural groups. The growth of oil revenues during the 1970's helps to explain Indonesia's shift from extraction to subsidization, for example, and the growth of subsidies in Malaysia. Likewise, over the same period the deterioration of the Philippine economy impelled the withdrawal of heavily subsidized credit for rice producers.

Political forces are nevertheless essential both in explaining agrarian strategies and in analyzing their implications for differentiation of rural society and economy. Prominent among these is rural dissent and the ways in which states in the region have sought to deal with it. The subordinate classes of rural society in Southeast Asia have not always remained passive in the face of exploitation. Nor have they awaited their transformation into a proletariat in order to express their dissatisfaction. Over the past few decades, all four states have at various times been confronted with organized rural resistance that represented a challenge to the structure of state power. Moreover, agrarian dissent has arisen as much or more in the major rice-producing areas as in the large-scale, wage-labor-based plantation sectors.

As discussed more fully in the country introductions and in the essay by Hart (Chapter 2), each of the four states has attempted to deal with the conflicting imperatives of ensuring rice procurement and maintaining agrarian control through a strategy of patronizing dominant rural groups. The character and composition of dominant rural groups and their relationship to the state vary in ways that reflect the historically specific differences described in the case studies. In all four cases, however, the position of dominant rural groups vis-à-vis the larger system both influences and is affected by their relationships with small peasants and landless laborers.

The case studies document different ways in which these forces are playing themselves out in particular locations in main rice-growing regions in which both technology and commoditization are highly developed. Moreover, all are based on restudies of a particular village or area and are therefore able to provide some degree of historical perspective.

It is important to bear in mind, however, that the case studies were not part of a grand research design. Rather, they represent a selection of local-level studies that were available at the time this project was conceived. Consequently they vary in approach, method, and coverage. Some focus on quantitative evidence (Lim; White and Wiradi), some give primary emphasis to political forces (Turton), while others attempt to incorporate quantitative and qualitative evidence with varying degrees of comprehensiveness (Anan, Banzon-Bautista, Fegan, Hüsken, Ikmal). Despite these variations in approach and scope, each study adds important insights to our understanding of differentiation.

We begin our case studies with Thailand because Turton's analysis of local powers (Chapter 4) plays a bridging role between the more general essays by White and Hart and the village-level studies which follow. Focusing on the exercise of power at different levels primarily in areas of northern and northeastern Thailand, Turton suggests one approach to the conceptual and methodological problems of linking the state and social processes.

Anan's study of a northern Thai village (Chapter 5) describes in rich detail the transformation of agrarian labor arrangements following the state's clamping down on peasant rebellions of the mid-1970's and the intensification of commercial rice production. He shows how sharecropping contracts are being renegotiated and adapted to allow the landlord greater control over the tenant's deployment of labor and nonlabor inputs, while also appropriating the incremental output. Although a source of some tension and resentment, these contracts provide tenants with a secure income in the face of uncertain labor market conditions. Workers excluded from these arrangements began engaging in collective action. The evidence provided in Anan's study complements Turton's emphasis on local powers and their linkages with the state.

Both Philippine studies (Chapters 7 and 8) are set in Central Luzon. Banzon-Bautista identifies a variety of links between the village and political and economic forces in the national and world economy. Following forty years of agrarian conflict in the region, including a full-scale armed rebellion in the 1940's, the state under Marcos intervened in 1972 to convert all sharecroppers to fixed-rent tenants, set low rents, and granted tenants of large owners the right to buy their farms on installments. But local landless people went further, occupying and dividing idle land of the absentee landowner while local farmers made a communal fishpond of her irrigation reservoir. Although labor, capital, inputs, and the product are market commodities, and a covert market in land persists, smallholders have not been displaced. Putting aside land gained by occupation, the average operated farm size has declined due to illegal division of farms among heirs. There is no evidence of kulak or capitalist reaccumulation of land, despite a covert market in which farmholders sell or mortgage land rights to get capital for nonfarm enterprises or raise bribes and agency fees to obtain exit papers and construction jobs in Arabian Gulf countries. Nonfarm savings, especially from labor abroad, are the major source of funds to purchase farm rights and capital in the form of agricultural machinery (which is rented out), vehicles, and small businesses. Low returns and high risk in rice agriculture encourage peasants to invest in ways of diversifying household income sources, including investment in the education of children.

Fegan's village study (Chapter 8) shows that mobile capital is not attracted to investment in rice, a crop with high natural risk plus low returns in Central Luzon at the time of the study. A combination of urban-biased government policies and world market trends have made the real price of rice fall in relation to all other prices. Under these conditions smallholders are protected from capitalist competition for land by the same forces that keep them poor as farmers. Where possible, mobile capital has left the rice monocrop area, but town and village capitalists committed to the rural economy have found profitable niches, trading in inputs and rice, moneylending, and hiring out machine services. The most successful accumulators trade across linked markets, lending money for farm expenses, machinery hire, and consumption against the security of the crop. They require the borrower to use their threshing machine, which then acts as a vehicle of debt collection. As in Banzon-Bautista's study, all households pursue a strategy of diversifying household sources of income, including investment to secure migrant labor contracts abroad, in nonfarm small businesses, and the education of children for urban professions.

Both the Malaysian studies (Chapters 10 and 11) examine the consequences of massive technological change in the Muda Irrigation Scheme in the northwest state of Kedah, which exemplifies the heavy state subsidies to rice producers. Lim (Chapter 10) shows how the small rice peasant owners were entrenched in the first phase of technological change, characterized by sharp increases in labor demand, although many tenants were displaced. Despite the predominance of small farms in Muda, a few very large-scale producers do exist. Ikmal (Chapter 11) documents the strategies of these large farmers and describes the enormous influence they exercise vis-à-vis the small peasantry. With the spread of combine harvesters since the late 1970's, Muda has now become the most heavily mechanized area of rice production in Southeast Asia. Ikmal shows how a group of large farmers has been diversifying into combine ownership and points out the ways in which they are trying to link combine services with credit.

White and Wiradi (Chapter 13) analyze agrarian changes in nine villages in different rice-growing regions of Java, using Agro Economic Survey data from 1971 and 1981. While there is great variation in the forms of agrarian labor relations and the degree of differentiation among these villages, a general picture emerges of increasing landlessness accompanied by growth in average farm size despite population growth and by a shift in the proportional division of shares of growing agricultural incomes in favor of farmers compared to hired labor. A variety of rationalizations in labor process, including both laborsaving technologies and exclusionary labor arrangements, underlie this general shift. Although these conclusions at first sight suggest the emergence of opposing "commercial farmer" and "landless agricultural laborer" classes at either end of a relatively declining mass of small-peasant households, the authors warn that such conclusions based on changing production relations in agriculture alone are misleading as a characterization of *rural* classes or class relations in these villages, because nonfarm incomes now represent almost two-thirds of total income. White and Wiradi contrast the "accumulation" and "survival" nonfarm strategies of agrarian elites and the landless and near-landless—patterns which call for models of agrarian differentiation in which the phenomenon of "part-time" farming at all levels of the agrarian structure is accorded a more central focus.

Hüsken's village study (Chapter 14), based on extended field research and historical work in Pati (northern Central Java), documents in some detail the changing relations between landed village elites and landless or near-landless wage workers and share tenants, focusing on mechanisms of surplus extraction. Commercialization and Green Revolution production have been accompanied by growth in the relative dominance of share tenancy compared to wage labor and by new forms of sharecropping contracts together with other forms of exclusionary labor arrangements which have allowed landowners to claim an increasing proportion of the harvest. The various economic activities and relationships in which these elites are involved allow Hüsken to describe them only as "hesitant" capitalists, commanding an increasing surplus but shrinking at heavy involvement in capitalist investment. The earlier history of this region underlines the fragile nature of the currently prosperous conditions of its "feather-bedded" elites, which depend partly on direct state patronage and more generally on an economy buoyed at the time by oil revenues but vulnerable to periodic recessions.

Despite the diversity of approaches and methods, there are several important—and often related—themes that cut across the eight local-level case studies in this volume:

1. The persistence of petty production in a variety of different circumstances and through multiple and changing mechanisms.
2. The adaptation of supposedly "precapitalist" labor-tying arrangements to allow the landlord-employer greater control over a comparatively "privileged" group of workers.
3. A process of diversification into nonagricultural activities which is particularly marked among the richest and poorest.

The persistence of both petty production and labor-tying has been observed in many other settings, provoking heated debate over the causes and consequences of these phenomena. By providing comparative evidence on variations and similarities among institutional arrangements occurring in different settings, the case studies help to clarify several important issues. For example, both the variations and changes in the mechanisms through which petty producers persist sound a strong warning against efforts to develop a general theory of the persistence of petty production. One can, nonetheless, discern important similarities in the logic of certain institutional arrangements that recur in different contexts, a notable example being the way in which labor-tying arrangements operate according to exclusionary principles.

The internal logic of a particular mechanism does not, however, provide a sufficient basis for understanding either its emergence or its implications (Hart, Chapter 2). Rather, labor-tying and the mechanisms through which petty production persist are a reflection of larger sets of political and economic conditions, as well as a source of change—often unintended—in these conditions. These processes can only be understood in the context of historically specific structures of state power.

Although we have gone part of the way in illustrating the centrality of the exercise of power at different levels to understanding agrarian change, much remains to be done. In particular, there is need for a clearer understanding of how relations between genders and generations are both shaped by and act upon the larger political-economic system.

A number of studies in this book and elsewhere (e.g., Heyzer 1985) document how poor rural women experience the harshest effects of technological change and commercialization. Changing relations between genders and generations are not just a consequence of rural change, however, but are almost certainly a key element in specifying the mechanisms through which change takes place. That culturally constructed relations between generations shape mechanisms of accumulation—and hence processes of structural change—emerges clearly from Ikmal's study in Chapter 11.

Recent efforts to conceptualize gender relations and "the household" in relation to larger processes of structural change are most fully developed in the literature on Africa, where the enormous variety and complexity of organizational forms impels attention to culturally specific relations between genders and generations (Guyer and Peters 1987). As noted by White (Chapter 1), recent conceptual advances reject the notion of the household as a solidary unit and focus instead on partly overlapping units of production, consumption, and accumulation whose organization can vary and change. This conceptual approach is closely compatible with the distinction that we draw between mechanisms and processes of change and with our emphasis on the exercise of power at different levels of society.

The need for a better understanding of relations between genders and generations is clearly called for in the analysis of diversification which, particularly among the poor, often entails shifts in the division of labor as well as complex patterns of migration and labor circulation that both reflect and shape national and international patterns of accumulation. A central challenge of future research will be to link the analysis of changing relations between genders and generations with emerging processes of differentiation and class formation.

References

Bernstein, H. (1979). "Concepts for the Analysis of Contemporary Peasantries." *Journal of Peasant Studies* 6:431–443.

Bray, F. (1983). "Patterns of Evolution in Rice-Growing Societies." *Journal of Peasant Studies* 11:3–33.

——— (1986). *The Rice Economies*. Oxford: Basil Blackwell.

Chevalier, J. (1982). *Civilization and the Stolen Gift: Capital, Kin and Cult in Northern Peru*. Toronto: University of Toronto Press.

Friedmann, H. (1980). "Household Production and National Economy: Concepts for the Analysis of Agrarian Formations." *Journal of Peasant Studies* 7:158–184.

Geertz, C. (1963) *Agricultural Involution: The Processes of Ecological Change in Indonesia*. Berkeley: University of California Press.

Guyer, J., and P. Peters (1987). "Introduction to 'Conceptualizing the Household: Issues of Theory, Method and Application.'" *Development and Change* 18:197–214.

Hayami, Y., and M. Kikuchi (1982). *Asian Village Economy at the Crossroads*. Baltimore: Johns Hopkins Press.

Heyzer, N. (1985). *Working Women in Southeast Asia: Development, Subordination and Emancipation*. Philadelphia: Open University Press.

Kahn, J. (1981). "The Social Context of Technological Change in Four Malaysian Villages." *Man* 16:542–562.

——— (1982). "From Peasants to Petty Commodity Producers in Southeast Asia." *Bulletin of Concerned Asian Scholars* 14:3–15.

Lehmann, D. (1982). "After Chayanov and Lenin: New Paths of Agrarian Capitalism." *Journal of Development Economics* 11:133–161.

Palmer, I. (1977). *The New Rice in Indonesia*. Geneva: United Nations Research Institute for Social Development.

Smith, C. (1984). "Forms of Production in Practice: Fresh Approaches to Simple Commodity Production." *Journal of Peasant Studies* 11:201–221.

Part One

Analyzing Agrarian Change: Issues and Problems

Chapter One

Problems in the Empirical Analysis of Agrarian Differentiation

BENJAMIN WHITE

> *The main trends of peasant differentiation are one thing: the forms it assumes, depending on the different local conditions, are another.* (Lenin 1976:145)

This chapter will focus on some problems in the empirical analysis of agrarian differentiation processes. Many of these problems are of a general nature, but some are of special relevance to Asian rice-growing societies such as those from which the case studies in this volume are drawn. All the situations described in the eight local case studies (and in the four country or regional introductions which accompany them) confront us with situations in which it is certainly possible and useful to speak of local processes of social-economic differentiation within a global capitalist context, but those processes cannot easily be equated with the straightforward or 'classical' model of differentiation as first systematically outlined in Lenin's analysis of the process of capitalist development in Russian agriculture at the turn of the century (Lenin 1960). 'Capitalist farmers' when present at all are a minority of farmers and rarely appear in pure forms, being involved also in a variety of noncapitalist relations with direct producers; the same could be said of the 'landless proletariat,' and while a part of agricultural production—

sometimes a substantial part—is realized through the application of wage labor on larger farms, alongside these farms we find larger numbers of the small-peasant units called in the classical literature 'labor farms' (Chayanov 1966; Kautsky 1899).

Similar conditions also confront researchers in most other regions of the Third World and indeed in many parts of the First. Decades of relatively intensive research have produced much debate, but less in the way of firm conclusions about the main tendencies of social and economic change or the specific forms it has taken in specific countries and regions. These problems derive partly from difficulties and disagreements about the specific issues deserving emphasis in field research; even if there were agreement on these questions, however, there would remain many problems of a purely practical nature in finding appropriate means of collecting and analyzing the necessary empirical information to throw light on them.

We will first examine some aspects of the first problem, at the same time touching on the question of the relevance of 'classical' models and debates on agrarian differentiation to contemporary (and, particularly, contemporary Southeast Asian rice-growing) societies. Before beginning, the reader should be warned that this chapter is not a place to look for answers to the various questions raised. Theoretical literature on differentiation processes and on the nature of contemporary peasantries generally—already much too large to attempt to review here—is currently in a state of some confusion (cf. editors' introduction) and perhaps likely to remain so for some time; this means among other things that there are advantages to strategies involving flexibility and openness rather than rigid adherence to a standard framework in empirical investigations, and the purpose of this chapter is mainly to argue for such flexibility.

Research 'method' involves the selection of general approaches and conceptual frameworks which point to relevant issues for research on processes of agrarian change, as well as the selection of specific procedures for data collection and analysis to throw light on those issues. Questions of research method in fact begin from starting points even further removed from problems of mere technique, since approaches to research are also determined by the broader strategic purposes of inquiry, which in turn are closely related (whether explicitly or implicitly, even consciously or unconsciously) to its specific political context. One crude contemporary example in 'agrarian differentiation' research is the analysis of data on trends in landlessness and peasant revolutions in order to construct an 'index of rural instability' for explicitly political purposes (Prosterman 1976)—in this case, as a basis for promoting counterrevolutionary land reform programs.

Much of the current confusion concerning the relevance of the so-called classical European models and debates on peasant differentiation appears to have ignored this dimension. It is therefore helpful to recall that classical European discussions on the differentiation of the peasantry in the late nineteenth and early twentieth centuries were undertaken for more or less immediate (and usually for explicit) political purposes, in relation to the 'agrarian questions' of the times and places in which they were conducted. These 'questions' were not posed out of neutral curiosity about the changing structure of rural society, but as strategic questions concerning the potential economic and political role of various sections of agrarian society, or of the peasantry as a whole, in the advance towards capitalist or socialist forms of development (Brass 1984; Hussain and Tribe 1983). These strategic questions provided for authors like Lenin and Kautsky, and others in different camps like Chayanov, the criteria of relevance in specific empirical issues for research; this is what led them into mundane questions of a technical nature, none of which are of any great interest in themselves, such as the differences between landless, one-horse, and allotment-holding 'proletarians' (Lenin), the comparative 'efficiency' of small and large farm enterprises (Kautsky), or the changing consumer/worker ratios at different stages in peasant family life-cycles (Chayanov).

These contextual aspects of classical differentiation literature seem to have been poorly understood by many subsequent researchers, as reflected in the wholesale transplantation of classical models (often in their most simplistic and rigid form) to quite different contexts, as if there were to be found in the works of Lenin, Kautsky, Chayanov, or other classical forebears a set of alternative 'research guides.' These remarks are not intended to imply that classical analyses and debates from Europe have no relevance to contemporary research in Asia. However, a better understanding of the links between the national political context, the strategic purposes, and the concrete focus of earlier investigations should free us from any notion that one or other model is the 'correct' one, to be faithfully replicated. There *is* no universal or all-purpose 'agrarian question' awaiting investigation, nor is there any universal form of 'agrarian differentiation' (hence the quotation which introduces this chapter); the search for all-purpose definitions of these notions, or a universally right way to analyze them in particular contexts, is as fruitless and unnecessary an exercise as the search for universal definitions of the 'peasantry' who are supposed to constitute the 'question' and to undergo or resist the 'differentiation' process (cf. Hussain and Tribe 1983:ch. 10).

Having said this, it is necessary to note that classical analyses have much to offer in the way of specific approaches and insights which are of relevance to contemporary situations; this relevance, however, is to be explored rather than assumed or imposed. The empirical analysis and interpretation of agrarian changes in specific situations demand a degree of flexibility on the part of researchers rather different from the kind of thinking involved at a high level of abstraction on the general tendencies or 'laws of motion' of capitalist development. Paradoxically, the classical authors we have mentioned were themselves more flexible in this regard than many subsequent researchers who have drawn on their work. Both Lenin and Kautsky, for example, make rather explicit distinctions between general or underlying 'trends' or 'tendencies' and the concrete forms of agrarian change apparent in specific situations. Much of Kautsky's work is devoted to discussing 'how particular factors at work in capitalist economies hinder realization of tendencies,' such as the factors which prevent large farms from replacing less 'efficient' small farms (cf. Hussain and Tribe 1983:105). Again, neither Lenin nor Kautsky seemed to feel at all uncomfortable in speaking of a process of rural 'proletarianization' while insisting at the same time that this does not at all require complete dispossession from land; rather, it requires a sufficient degree of inequality in access to land and other production resources to leave large numbers of 'peasant' households in possession of farms incapable of providing a livelihood and therefore propelling one or more household members partly or completely into the agricultural or industrial wage-labor market (Kautsky 1899:ch. 8). Such discussions are clearly of great relevance to the understanding of the 'survival' of small-peasant farm households, and the contradictory or ambiguous 'classness' of their members, as several of the case studies in this volume indicate.

Flexibility and openness in investigations of concrete situations, in contrast with the abstract rigor of theoretical formulations, seems a natural and healthy consequence of the recognition that economic and social changes occur in actual societies with their own configurations of political forces at local and higher levels, with all kinds of complex and sometimes conflicting processes at work both within and beyond the village, whose interaction with general 'tendencies' results in specific patterns of differentiation. When such flexibility is lacking we are left with less useful analyses—such as those which hold up some empirical data against a rigid version of a supposed Leninist or Marxist model involving the emergence of two and only two opposing classes, note that they have failed to materialize or to extinguish the others, and conclude that the model 'does not work' for this region, for that country, or for the Third World.

Nor is it any more productive to set against each other the two 'opposing' models of Lenin or Chayanov, again in the most extreme and rigid form of each. The analyses of both these authors in fact provide many insights which are relatively independent of their general models and not incompatible with each other, but rather highlight different processes which may be simultaneously at work in agrarian societies. Although we may easily reject the essentialist model of peasant economy on which Chayanov's work is based, Marxist analysis has hardly come to grips at all with many of the other issues raised by Chayanov (Gledhill, 1985:51). The analysis of peasant family demographic cycles, for example, helps us to identify mechanisms of cyclical (nonpermanent) mobility which are responsible for part of the observed distribution of owned land or farm sizes at a given time; the Leninist focus on the mechanisms of 'cumulation of advantages and disadvantages' aims to identify and understand the more permanent, noncyclical component of these inequalities and their changes over time. Few analyses have even attempted to disentangle these coexisting processes of cyclical and permanent mobility, which is indeed a technically complex task (cf. Shanin 1982; Deere and de Janvry 1979; Cox 1986). Many studies continue to present us with statistics on land concentration and landlessness as if these were adequate indicators of rural class differentiation, unwittingly including among the 'landless class' young couples waiting to inherit land from their landed parents or former proprietors who have made over land to their children, not to speak of the local policeman or family-planning fieldworker assigned from the capital city and other anomalous 'landless' households, who may never experience wage labor or class exploitation, let alone join the agrarian struggles of the landless.

Agrarian or rural 'differentiation,' as the term implies, is a dynamic process involving the emergence or sharpening of 'differences' within the rural population, but it does not itself consist of (and in some cases, at least in the short term, may not even involve) increasing income inequalities. It is not about whether some peasants become richer than others but about the changing kinds of relations between them (or between peasants and nonpeasants, including extrarural groups) in the context of the development of commodity relations in rural economy. The changes involved in differentiation processes are thus essentially qualitative rather than quantitative, although of course they may be quantitatively measurable; they involve changes in the form or at least in the function (see below) of production relations; they can occur in contexts of a stagnating, expanding, or declining rural economy, although the forms which differentiation takes in each case would probably be different. In most of the Southeast Asian case studies in this volume, the recent context of differentiation has been one of agrarian growth through Green Revolutions of varying success, although recent experience in the Philippines offers an important counterexample.

Differentiation thus involves a cumulative and permanent (i.e., noncyclical, which is not to say that it is never reversible) process of change in the ways in which different groups in rural society—and some outside it—gain access to the products of their own or others' labor, based on their differential control over production resources and often, but not always, on increasing inequalities in access to land. We might, for example, encounter situations in which landownership and farm size distributions remain fairly stable but agrarian surpluses derived from productivity growth are devoted to the acquisition of other production resources (tractors, water pumps, agro-processing or rural industry plant, means of transport, etc.)—all of which function to divert shares in the product of existing activities from one group to another (for example, from manual laborers to the owners of tractors) or channel incomes from new activities to particular groups, thus setting in motion the process of 'cumulation of advantages and disadvantages.'

The focus of investigation of these changing relationships concerns the mechanisms of transfer or extraction of surplus generated in rural economy, as may be seen in the work of many authors who do not necessarily share a general theoretical perspective but consider these relationships as the fundamental determinants of agrarian structures. Wolf, for example, considers the appropriation of surplus a defining characteristic of peasantries and regards the specific forms of extraction as a defining characteristic of different types of peasantry (Wolf 1963). Deere and de Janvry's 'conceptual framework for the empirical analysis of peasants' (Deere and de Janvry 1979) also rests on the identification of different mechanisms of surplus extraction in the production and circulation processes in which peasant household members may be involved: rent in labor service, cash, or kind; surplus value in wage relations; terms of trade in the sale and purchase of commodities; taxes, and interest. Ghose (1983:8) builds a framework for the analysis of developing agrarian economies around the notion that "the sources of dynamism (if any) of agrarian systems in developing countries must be sought in the sphere of relationships between the laboring class and the nonlaboring recipients of a share of the produce of the land, i.e., outside the domain of the independent peasantry."

These authors then share (although they may reject other elements in Marxist theory) a basic analytical starting point of identifying the economic foundation of social (in this case, agrarian) structures and their dynamic in "the specific economic form in which unpaid surplus labor is pumped out of the direct producers, [which] determines the relationship of domination and servitude. . . . On this is based the entire configuration of the economic community arising from the actual relations of production, and hence also its specific political form" (Marx 1981:927). While this may seem a relatively straightforward guiding principle for empirical research, in actual situations it may become extremely complex. (For a recent attempt to operationalize the 'surplus' criterion of agrarian class status, see Athreya et al. 1987.) In nearly every concrete example one can think of, surplus is 'pumped out' of direct producers in not one but many different forms. Moreover, not only do different forms of surplus extraction coexist, but the functions of these specific forms (in relation to the dynamics of accumulation by nonproducers) may change while the form remains unaltered; this may be the case, for example, with various forms of share tenancy and labor-tying practices mentioned in several of the case studies. Different forms of labor and of surplus extraction are found not only within the community but also within one person. A 'marginal peasant' male may be involved in tenancy, debt, and wage relations with the same individual or a number of different persons. As a petty commodity producer on his owned or tenanted plot, or in nonfarm 'own-account' production, and at the same time as a consumer of urban products, he may be involved in relations of unequal exchange. He may be subjected to formal taxes and informal exactions by various village and state officials and local strongmen. (The latter is a mechanism of extraction which, perhaps because of its coercive rather than economic form, is omitted from Deere and de Janvry's framework and may not derive from any specific form of labor but simply from the relative power and powerlessness of the parties involved.) When the unit of analysis is raised from the individual to the household (and most analyses do use 'households' as their basic unit of analysis, perhaps wrongly), there may be still more forms of extrahousehold surplus extraction mechanisms in which subordinate (female and juvenile) household members are involved.

The 'household' as a unit of analysis is an extremely problematic concept. Long used as a kind of catch-all minimal social-economic unit, it is now becoming increasingly recognized that we should really think more in terms of separately and carefully defined (and only partly overlapping) units of production, consumption, accumulation, etc., sometimes to the extent of abandoning the concept of 'household' altogether. (See the many recent collections of studies in Guyer and Peters 1987; Long 1984; Netting et al. 1984; and Smith et al. 1984.) Certainly notions of household must include the possibility of surplus extraction mechanisms between household members based on hierarchies of age and gender, such as the process which Cook has recently termed 'endo-familial accumulation' based on unpaid family labor (Cook 1984:29) and their relationship to wider processes of accumulation and social differentiation—a matter which deserves much more explicit attention in the case studies.

Besides this, it should already be clear that although 'community' or 'village' (the focus of 'village studies') may be easy to define as a social unit, it is of little use as a unit confining analysis. A significant source of the 'survival' of sublivelihood peasant farms, for example, may be found in the labor migration of rural household members to urban centers or even across national boundaries. Banzon-Bautista's study from a Philippine village (Chapter 7) shows that the majority of all gains and losses in land and changes in incomes over a seven-year period were directly related in some way to the 'Saudi connection.' Both labor and capital mobility across village boundaries and between rural and urban and agricultural and nonagricultural sectors mean that we would obtain a very misleading picture of rural differentiation, and of the political component of rural 'classness,' by confining our analysis only to intra-village relationships or to production relations in agriculture.

In trying to characterize forms of differentiation in Asian rice-growing societies in some general way, various authors have suggested that those who expect to find wage-labor forms and some variant of 'two-class polarization' as the inevitable consequence of commodity production and capital penetration are looking for the wrong thing. Bray has recently argued that paradigms of differentiation based on the experience of Northwest European agricultural systems have failed to understand the relative peculiarity of those systems, whose dynamic has been historically conditioned by the 'superior performance of large, centrally managed units of production' so that the immediate consequence of the expansion of productive forces was 'the polarization of rural society into farmer-managers and wage-laborers' (Bray 1983:8). Asian wet-rice cultivation, on the other hand, has enormous potential for increasing land productivity in relatively cheap and scale-neutral ways, involving added inputs of manual labor rather than capital. Under these conditions a smallholder or tenant farmer is in just as good a position to raise land productivity as a wealthy landlord (ibid.:11); landlords gain little by evicting tenants to run large centrally managed estates, and hence we see little evidence of 'consolidation of holdings' (ibid.:13) or polarization. Besides viewing these developments in a long-term historical perspective, Bray considers them to apply equally to 'modern' Asian rice farming: "Even today, wet-rice economies prove remarkably resistant to the transition to capitalist farming. . . . Despite the commercialization of rice farming, capitalist relations of production have not resulted: we do not find the polarization of rural society into large farmer-operators and landless wage-laborers, and the basic unit of production remains the family farm" (ibid.:23).

Wet rice is of course not the only non-European crop regime in which polarization in pure form has not emerged, nor did polarization occur in all of Northwest European agriculture; and we may wonder whether Bray is correct in focusing on the characteristics of the crop itself as the cause of its relative absence in Asian wet-rice-growing societies. Besides that, large mechanized wage-labor rice farms do exist, and not only in the United States or Australia (cf. the study of large farms in the Muda region by Muhammad Ikmal Said, Chapter 11) even though in Asia they coexist with smaller owned or tenanted 'family farms.'

Another recent contribution focusing specifically on Southeast Asian wet-rice village economies is Hayami and Kikuchi's counterposition of what they call the 'two opposite directions' of 'polarization' and 'stratification.' The first is the conventional two-class polarization process in which "personal relations between employers and employees (or landlords and tenants) which have been guided in village communities by traditional moral principles such as mutual help and income-sharing will be replaced by impersonal market relations" (Hayami and Kikuchi 1981:60). The second is the process of "increasing class differentiation in a continuous spectrum ranging from landless laborers to noncultivating landlords, while the social mode of traditional communities is maintained. . . . People are tied to one another in multi-stranded personalized relations, and all community members have some claims to the output of land. . . . In the peasant stratification case semi-subsistence peasants will survive, although the majority of them may become poorer with smaller farms to cultivate" (ibid.).

The 'peasant stratification' case thus also depicts a process of 'cumulation of advantages and disadvantages' similar to Lenin's 'differentiation of peasant agriculture from within.' It is not clear why the authors see these two processes as representing two opposite directions or regard Southeast Asian village economies as standing at the 'crossroads' between them (ibid.:62). Hayami and Kikuchi themselves seem ambivalent on this question, since they conclude with reference to 'forces which may combine to create the threshold from which the peasant stratification process evolves into the polarization process' (ibid.: 225)—suggesting that the two trends are not aptly depicted as opposite directions at a crossroads but rather (if traffic metaphors are appropriate) as successive stages on a single unidirectional route, with various forces at work to speed or delay the transition to the second. The progression envisaged is then not essentially different from that depicted by Lenin: first, 'differentiation from within' in which peasant groupings and the relations between them become more complex; second, the 'simplifying' process of polarization of rich, middle, poor peasants and landless workers into only two opposing classes linked by wage relations. There is also some reason to doubt whether the emphasis on 'personalized' vs. 'impersonal' relations of surplus labor appropriation is necessarily related to the breakdown or maintenance of traditional 'social modes' of income sharing, etc. Various of the case studies in this volume document the emergence or spread of personalized share tenancy or labor-tying relations which function precisely to exclude some community members from claims to the agricultural product (Hart, Chapter 2).

Kahn has argued (using examples from Malaysia, West Sumatra, and Java) that the frequent emphasis on 'microclass' differences within rice-growing villages leads to misleading characterizations of both the 'objective' and 'subjective' aspects of rural class formation. Differences between peasants—a form of social stratification 'more akin to precapitalist social hierarchy' (Kahn 1985:88)—pale into insignificance beside the differences between peasants as a whole and the various supra-village elite classes who extract surplus from them, often through noncapitalist mechanisms (Kahn 1983). This intervention, while making one important point, seems to have lost sight of another. Certainly traditional village-level analyses must be expanded to include relations of surplus extraction and political antagonism between peasants as a whole and nonpeasant groups at various levels, from regional and national to world system. Nevertheless, the currently popular focus on petty commodity producers and 'reconstituted peasantries' in relation to outside forces is in danger of neglecting the existence of large farmers and their role in both the economic and political dynamics of rural society. The relatively small absolute size of 'large' farms in the Southeast Asian context (they may, for example, be defined as those over 5 ha in the Philippines, 3 ha in Muda, or even 1 ha in the extreme case of Java) does not mean that class relationships between peasants do not exist, cannot be objectively defined, are not viewed by those involved in them as class relations, or do not become the focus of political struggles. Those who own or control these 'large' farms, as the case studies show, command substantial surpluses, are sellers of commodities produced mainly or entirely by the labor of others, and through their relative dominance in agricultural production (although they represent a minority of farms) and local power relations have access to many of the most lucrative nonagricultural opportunities, displaying in many ways the classical features of the kulak. As argued by Hart (Chapter 2) they have been incorporated as the clients and political allies of contemporary Southeast Asian states in the political control of the agrarian population, and a large part of the political struggles of small-peasant and landless men and women centers on the conditions of surplus labor appropriation by this group, regardless of other struggles in which 'peasants' as a whole may combine forces against 'classes defined outside the village economy' (Kahn 1985:79).

It is useful to make a distinction between the *process* of differentiation itself and various aspects of that process which we might call the *causes*, the *mechanisms*, and the *symptoms* or *indicators* of differentiation. Similarly, any analysis of rural differentiation processes in a specific place and time will have to encompass their *contexts* (regional, national, political, cultural, etc.) and also the *constraints* to differentiation (which may originate externally or internally and may affect the pace and form of differentiation).

Examples may help to make some of these distinctions clear. The *causes* of differentiation are generally sought in the penetration or expansion of commodity economy and therefore require the extension of analysis to supra-local, national, or global levels. They may involve, as they specifically affect rural producers, either externally induced shifts in demand for the products of the agricultural sector through domestic industrialization or the opening up of export markets or changes in the technological basis of agriculture.

The *process* of differentiation itself, as we have already suggested, concerns shifts in patterns of control over means of production and the accompanying social division of labor. As examples of the *mechanisms* through which such shifts occur we could include on the one side the resumption of tenanted land by its owners and a variety of other mechanisms of partial or total dispossession of land and other production resources; on the other side are the various alternative forms of disposition of agricultural surpluses by rural elites. External forces and historical political *contexts* can exert great influence on these mechanisms (or 'forms') of differentiation; we may mention, for example, Fegan's argument that national political-economic aspects of the Philippine rice industry made rice production itself relatively unprofitable, so that large landowners and rich peasants have not generally made efforts to gain or regain control over substantial areas of land since land reform but have diversified into more profitable areas of rural economy (Fegan, Chapter 8). Local historical experience of agrarian struggles may also exercise important influence on forms of differentiation; the recent history of agrarian struggles in Thailand, for example, has made some landowners prefer diversification and the leasing out of land in the face of uncertain labor supply or quiescence (Anan Ganjanapan, Chapter 5) just as the different history of agrarian conflict in Java now makes it politically feasible for large farmers to engage in rationalizations of labor process without significant mass opposition (Hüsken and White, Chapter 12). Such examples as these highlight the importance of *contexts* and *constraints*, which as suggested earlier make the empirical analysis of differentiation processes a rather different kind of exercise from the abstract formulation of 'general' tendencies.

The *symptoms* or *indicators* of differentiation include such features of rural social-economic structures as the distribution of owned and operated land; frequency and form of tenancy relations and the direction of land flows through tenancy between landownership groups; family-, exchange-, and hired-labor use; and investments and incomes of men and women in different groups or classes in different activities. Although perhaps stating the obvious, it is still useful to note that while studies of agrarian change generally aim to provide an understanding of processes and mechanisms of differentiation, the information gathered and presented in such studies is generally, or mainly, on indicators—i.e., it documents symptoms rather than processes although detailed local-level work can more easily combine this with information on mechanisms of change. This is true for contemporary researchers as much as it was for Lenin, Kautsky, or Chayanov (or the still more sophisticated analyses of Kritsman and his school in the 1920s; cf. Cox 1986), whose works are full of tables documenting farm size distribution, wage-labor use, ownership of draft animals and implements, sideline occupations, and so on. The arguments made about differentiation processes (or their absence) and their causes, then, are based on a combination of derivations from preexisting frameworks or theories, on the one hand, and inference from observed indicators or symptoms on the other, rarely supplemented by any direct documentation on processes and mechanisms in the case of the authors just mentioned, although many of the local-level case studies in this volume have achieved more in this regard.

This is one main cause of the possibility of contradictory interpretations of data on 'symptoms' which may not themselves be substantially different—as we may see, for example, in the totally opposing emphases placed by Lenin and Chayanov on data which showed both of them that 'capitalist farms' were a minority of all farm enterprises in Russia or the many articles of Lenin castigating other authors for the tendentious ways in which farm-size data in various other European countries are interpreted. (For the latter see Lenin 1976.) For Lenin, the fact that the 'peasant bourgeoisie' constituted a minority ('probably less than one-fifth') of all households did not mean that they could be ignored in analysis: "As to their weight in the sum-total of peasant farming, in the total quantity of means of production belonging to the peasantry, in the total amount of produce raised by the peasantry, the peasant bourgeoisie are undoubtedly predominant. They are the masters of the contemporary countryside" (Lenin 1960:177). This led him in turn to focus analysis mainly on the relationships between this group and the land-poor and landless, by examining statistics on the 'upper 20%' and the 'lower

50%' of the peasant population. Chayanov began with a starting point not much different—that '90% of the total mass of peasant farms are pure family farms' (Chayanov 1966:112), i.e., one-tenth *were* wage-labor farms. He also noted that "simple, everyday observation of life in the countryside shows us elements of 'capitalist exploitation.' We suppose that, on the one hand, proletarianization of the countryside and, on the other, a certain development of capitalist production forms undoubtedly take place there, [but] as we are concerned with the labor farm the themes we have touched upon, despite their exceptionally intense and topical general economic interest, are quite to one side" (ibid.:245, 255–257). Chayanov therefore focused in his analytical work exclusively on the model of the 'family labor farm.'

The point of this contrast is to suggest that many contemporary analyses also tend to select their own restricted focus and thereby to simplify complex situations. In nearly all contemporary Asian situations, two quite different analytical focuses are justifiable. The first emphasizes the 'upper' and 'lower' groups, the processes giving rise to land concentration and capitalist farming on the one hand and (near) landlessness and proletarianization on the other, and the relations between these groups. This focus approximates the position once taken by Rudra in trying to characterize class relations in Indian agriculture: "In Indian agriculture today, one can distinguish only two classes . . . a class of big landowners and a class of agricultural laborers. There are members of the working population in agriculture who do not belong to those two classes. In our view, they do not constitute or belong to any class or classes" (Rudra 1978:1001). The second focus—equally justifiable but no less distorting if it excludes the other—would ask why so many millions of rural households manage to hang onto tiny farms and resist the process of polarization—in other words the familiar question of the tenacity of small-peasant households in contexts of overall capitalist penetration of the economy. In the analysis of most situations, frameworks are needed which do not restrict us to either focus but allow us to incorporate them both as opposing but coexisting 'tendencies' and focus precisely on the nature and implications of their coexistence.

Any simplistic model of differentiation processes is likely to be unfruitful for analysis of the dynamics of change in local-level agrarian structures, which will reflect at any time the operation of a large number of complex, simultaneous, and often conflicting processes, some of which cancel each other out over time while others do not. As already noted in the editors' introduction, the various studies collected in this volume did not set out with a common research design or a single theoretical model of agrarian differentiation for empirical investigation. The sheer complexity of the forms which differentiation takes and the forces which influence it makes their comprehensive analysis an almost impossible task for any single researcher or research team. Most individual studies can at best hope to illuminate some aspects of differentiation processes, and many such studies need to be available before any fruitful attempt at synthesis can be made. An open and flexible attitude among researchers as regards the focus, level, and techniques of investigation may also have advantages in view of the uncertainty in recent literature even about such basic matters as what 'agrarian differentiation' is and why it deserves attention, let alone more practical questions of how the understanding of differentiation processes in specific contexts can be most usefully advanced through field research. Easy generalizations about agrarian change and rural class formation processes, drawn more from standard models than from careful empirical research, are both intellectually and politically irresponsible, and the case studies in this volume have recognized the need for caution.

References

Athreya, V., G. Boklin, G. Djurfeldt, and S. Lindberg (1987). 'Identification of agrarian classes: a methodological essay with empirical material from South India.' *Journal of Peasant Studies* 14, no. 2:147–190.

Brass, T. (1984). 'Permanent transition or permanent revolution: peasants, proletarians and politics.' *Journal of Peasant Studies* 11, no. 3:108–112.

Bray, F. (1983). 'Patterns of evolution in rice-growing societies.' *Journal of Peasant Studies* 11, no. 1:3–33.

Chayanov, A. (1966). 'Peasant farm organization.' In D. Thomas, B. Kerblay, and R. Smith, eds., *A.V. Chayanov on the Theory of Peasant Economy.* Homewood, Illinois: American Economic Association.

Cook, S. (1984). "Peasant economy, rural industry and capitalist development in the Oaxaca Valley, Mexico.' *Journal of Peasant Studies* 12, no. 1:3–40.

Cox, T. (1986). *Peasants, Class, and Capitalism, The Rural Research of L. M. Kritsman and His School.* Oxford: Clarendon Press.

Deere, C., and A. de Janvry (1979). 'A conceptual framework for the empirical analysis of peasants.' *American Journal of Agricultural Economics* 61, no. 4:601–611.

——— (1981). 'Demographic and social differentiation among northern Peruvian peasants.' *Journal of Peasant Studies* 8, no. 3:335–366.
Ghose, A. (1983). 'Agrarian reform in developing countries: issues of theory and problems of practice.' In A. Ghose, ed., *Agrarian Reform in Developing Countries*. London: Croom Helm.
Gledhill, J. (1985). 'The peasantry in history: some notes on Latin American research.' *Critique of Anthropology* 5, no. 1:33–56.
Guyer, J., and P. Peters, eds. (1987). 'Conceptualising the household: issues of theory and policy in Africa.' *Development and Change*, special issue, 18, no. 2.
Hayami, Y., and M. Kikuchi (1981). *Asian Village Economy at the Crossroads*. Tokyo: Tokyo University Press.
Hussain, A., and K. Tribe (1983). *Marxism and the Agrarian Question*. London: Macmillan.
Kahn, J. (1983). Review of G. Hansen, ed., *Agricultural and Rural Development in Indonesia*. *Journal of Peasant Studies* 10, no. 4:280–283.
——— (1985). 'Indonesia after the demise of involution: critique of a debate.' *Critique of Anthropology* 5, no. 1:69–96.
Kautsky, K. (1899). *Die Agrarfrage*. Stuttgart: Dietz.
Lenin, V. (1960). *The Development of Capitalism in Russia*. Vol. 3 of *Collected Works*. London: Lawrence & Wishart.
——— (1976). *The Agrarian Question and the 'Critics of Marx.'* Moscow: Progress Publishers.
Long, N., ed. (1984). *Family and Work in Rural Societies: Perspectives on Non-Wage Labor*. London: Tavistock.
Marx, K. (1981). *Capital*. Vol. 3. London: Penguin Books.
Netting, R., R. Wilk, and E. Arnould, eds. (1984). *Households: Comparative and Historical Studies of the Domestic Group*. Berkeley: University of California Press.
Prosterman, R. (1976). 'IRI: a simplified predictive index of rural instability.' *Comparative Politics* 8, no. 3:339–354.
Rudra, A. (1978). 'Class relations in Indian agriculture.' *Economic and Political Weekly* 13, nos. 22–24:916–923, 963–968, 998–1004.
Shanin, T. (1982). 'Polarization and cyclical mobility: the Russian debate on the differentiation of the peasantry.' In J. Harriss, ed., *Rural Development*. London: Hutchinson University Library.
Smith, J., I. Wallerstein, and H-D. Evers, eds. (1984). *Households and the World Economy*. London: Sage.
Wolf, E. (1966). *Peasants*. Englewood Cliffs: Prentice-Hall.

Chapter Two

Agrarian Change in the Context of State Patronage

GILLIAN HART

Despite the profoundly political nature of the "agrarian question" in any particular context (White, Chapter 1), most analyses of rural transformation and differentiation view the state as external to agrarian economic processes. This neglect of power and politics results in an almost exclusive focus on commercialization and technology as the main sources of rural change and portrays agrarian change as a unilinear process leading to a determinate outcome.

Comparative study of Southeast Asia dramatizes the perils of ignoring the state's role in rural differentiation. In much of Southeast Asia, patronage of dominant rural groups is an important means by which those who control the state pursue their complex and often conflicting agrarian interests, both within and beyond the rural sector. State patronage is central to understanding agrarian processes. It not only influences forms of extraction and accumulation, but also generates tensions and contradictions that constitute important sources of change and differentiation.

In concentrating attention on state patronage, I am not suggesting a peculiarly Southeast Asian "model" of rural differentiation. My purpose instead is to show how a historically specific analysis of the exercise of power at different levels of society enhances our understanding of rural differentiation. Indeed, the analytical value of comparative study of Indonesia, Malaysia, the Philippines, and Thailand derives partly from variations in state patronage that reflect important differences in power structures and the agrarian interests of the four states.

Although the particular forms of state patronage vary, they embody a set of broadly similar imperatives (discussed in the Introduction) to which different states in the region are subject. First, all four states have sought to contain or forestall the open expression of rural dissent that has characterized recent Southeast Asian agrarian history.[1] Second, they have been vitally concerned with ensuring rice production and procurement. Southeast Asian states have thus sought to maintain agrarian control while simultaneously transforming the productive base of rural society.

All four states have tried to deal with these conflicting interests by maintaining dominant rural groups—or a segment of them—as privileged yet dependent clients.[2] From the viewpoint of those who control the state, these rural elites are their agents in the countryside, helping to maintain control over the larger rural society and, in so doing, ensuring the political conditions of accumulation in the system as a whole. Privileged access to state resources is both a reward for cooperation and a means by which states have tried to ensure rice production and procurement.

Just as state patronage in its present forms has developed partly as a consequence of rural dissent in the past, so too it gives rise to new forms of conflict and tension. The central argument of this chapter is that state patronage contains the seeds of its own destruction and is likely over the long run to generate new threats to the structure of state power and hence new strategies of state intervention.

In developing this argument, I shall focus on how the linkages of dominant rural groups to the larger system both influence and are affected by their relations with subordinate classes. Drawing mainly on evidence from the case studies, I shall try to illustrate how local-level mechanisms of extraction and accumulation have an important political dimension that both reflects and alters the larger configuration of political-economic forces. The structure of the national economy and international economic forces are also essential to understanding the state's agrarian strategy and its contradictory consequences for rural differentiation.

State Patronage in Thailand and Indonesia: Unproductive Investment and Labor-Tying

From a purely macroeconomic point of view, the interests of the Thai and Indonesian states in the rice sector differ widely. As discussed more fully in the country introductions, the Thai state engages in net extraction from the rice sector, whereas the Indonesian state has provided increasing subsidies particularly in the period of expanding oil revenues. The political strategies of the Thai and Indonesian states are, however, quite similar: maintaining tight control in the countryside in order to circumvent agrarian mobilization that could be used to challenge the highly centralized structure of state power. The "floating mass" strategy, through which both states have sought to eliminate political mobilization of and by the poorer peasantry, has as its correlate the co-optation of dominant rural groups into the state machinery.

Rural elites become, in essence, political and economic agents of the state in the countryside and are co-opted into the larger structure of power as preferred but dependent clients. Their access to subsidized credit, inputs, licenses, guaranteed prices, and so forth stems not so much from their ability to sway agricultural policy in their favor through direct influence, but rather from the services that they render to larger centers of accumulation by helping to police the countryside. Although the aggregate flow of resources into the rice sector is more limited in Thailand than in Indonesia, these resources are channeled to dominant rural groups in rather similar ways.

These similarities help underscore the central importance of the state's efforts to ensure social control in shaping its agrarian strategy. At the same time, however, important differences exist in the nature and degree of direct state control over the rural elite. In the case of Java, the village upper stratum is increasingly being incorporated into the state machinery as minor functionaries (Hüsken and White, Chapter 12; Hart 1986a). Although technically there exists a system of election to local offices and a party political structure, in practice representatives of the state intervene actively to ensure allegiance to GOLKAR, the state electoral machine. Despite the coalescing of interests between the village upper stratum and supra-village officials through the formation of businesss alliances such as those described by Hüsken, the tightening of state controls at the local level has meant that this process of incorporation is increasingly taking place through official fiat.

The mechanisms through which the Thai village upper stratum is drawn into the supra-village sphere leave much more room to maneuver. The malleability of these alliances reflects the more unstable and fluid Thai political situation and contrasts quite sharply with the far more rigid local-level structures in Indonesia which are tied to a more authoritarian political system.

These differences reflect contrasts in the relationship of military and bureaucratic state structures to different branches of capital and the power struggles thereby engendered. In the case of Thailand, Turton points out that the weak development of a parliamentary and party political system at the national level strengthens the position of local powers and contributes to a considerable degree of fluidity in political alliances at the local level. Constraints on the development of a parliamentary system in turn reflect conflicts and compromises between the military and bureaucratic elites who control the state apparatus and urban-based capitalists (Girling 1985).

Although the military-bureaucratic state in Indonesia is similar in many ways to its Thai counterpart, the coalition in control of the state occupies a far more powerful position vis-à-vis domestic capitalists. Different factions within the bureaucracy engage in joint ventures with Chinese capitalists whose political position is highly precarious, while also seeking to define the conditions of existence of indigenous capitalists to prevent them from developing into an opposing political force (Robison 1978; Anderson 1983). The more heavily centralized structure of state power in Indonesia than in Thailand is, as we shall see, reflected in patterns of rural accumulation and labor control.

State Patronage and "Unproductive" Investment

Patterns of accumulation by Thai and Javanese rural elites that emerge from the case studies share a number of important similarities. The Thai and Javanese village upper strata are actively engaged in accumulation both within and beyond agriculture. Moreover, both Turton and Hüsken show how the dominant elite are directly involved in business alliances with supra-village authorities that sometimes also entail connections with Chinese merchants who are not directly involved in agriculture.

Despite rural elites' commitment to obviously capitalist forms of accumulation, the Thai and Javanese case studies also reveal very high levels of expenditure on luxury consumption and other apparently unproductive forms of resource deployment. Both the patterns of expenditure and the immediate reasons for them appear somewhat different, however. In the Thai case, Turton observes that a great deal of the expenditure on travel, social events, feasts, and so forth is in effect "geared to establishing and reproducing social relations with strategic superiors and subordinates in order to enhance political and economic position, to secure lucrative offices and contracts, to gain protection for illegal economic activities, [and] to accumulate political and economic clients." These activities are clearly designed to ensure the conditions of accumulation in the long run. Turton notes that even though some of the dominant elite have reached a point where they no longer depend on direct state handouts, they still rely on the coercive apparatus of the state and access to political office in order to maintain their position.

In the Javanese village studied by Hüsken, as well as in the village where I conducted fieldwork (Hart 1986a), it appears that the extent to which resources are devoted to maintaining social relations is far more limited than in the Thai case. The Javanese village elite entertain their superiors quite lavishly, and resources are probably invested in these upward connections in ways not very different from those observed by Turton in Thailand. Investment in relations with subordinates is almost certainly much more circumscribed, however. Turton describes how the Thai village upper strata are engaged in various forms of social and religious patronage at the village level through which they associate themselves with traditional forms of legitimacy that in turn have the ideological effect of mitigating conflict and resentment against inequality, as well as mobilizing political support. Such patterns are notably absent in Java. Although the exclusionary labor arrangements discussed more fully below do incorporate elements of "patronage," these are distinctively different from the direct investment in selective patronage relations with subordinate classes described by Turton. Unlike their Thai counterparts whose apparently unproductive forms of resource deployment are in practice geared towards long-run accumulation, the Javanese rural elite are described by Hüsken as "hesitant capitalists": "In the face of their suddenly increased surplus and well aware of its fragility they shrink at too heavy involvement in capitalist investments."

The key to explaining these differences lies in the mechanisms of state patronage that in turn reflect particular structures of state power. As discussed above, both the Thai and Indonesian states' efforts to maintain control over the countryside have entailed the co-optation of dominant rural groups, but the mechanisms through which these groups are incorporated into the state machinery are significantly different.

The comparatively high levels of investment in social relations with both superiors and subordinates described by Turton in Thailand reflect the far more fluid and malleable system of alliances at the local level through which the village upper strata are drawn into the supra-village sphere. These in turn derive from the more open conflicts at the national level between the military-bureaucratic elite and urban-based capitalists.

The more heavily centralized structure of state power in Indonesia has given rise to a far more tightly controlled system of incorporation, and developing a base of political support within rural society is far less important to achieving positions of power at the village level than in Thailand. As dominant rural groups are becoming minor state functionaries, their position is increasingly determined by more powerful agents of the state. To the extent that investment opportunities require official connections, licenses, contracts, and so forth, the hesitancy of rural elites to make a full commitment to capitalist accumulation observed by Hüsken may reflect the greater uncertainty of their position vis-à-vis the state, as well as the fluctuating macroeconomic conditions discussed by Hüsken and White.

Tied-Labor Arrangements

The relationship of dominant rural groups to the state is also central to understanding the forms and consequences of labor relations. The case studies by Anan Ganjanapan and Hüsken document in detail a particularly significant pattern that has recently been observed elsewhere in South and Southeast Asia—namely the adaptation and extension of complex tied-labor arrangements, especially by the largest employers.[3] Although the changes documented by both Hüsken and Anan appear to entail the transformation of sharecroppers into permanent workers, one should not label these adapted contracts "transitional forms" and presume that they simply represent points on a linear trajectory leading inexorably to single-stranded wage-labor arrangements. Apparently "transitional" contracts have been appearing, disappearing, and reappearing in Java since the nineteenth century (Hüsken and White, Chapter 12; Hart 1986a), and both Hüsken and Anan note that some contractual adaptations invoke much older institutional forms. Further, these relatively elaborate tied-labor arrangements generally coexist with far simpler and more casual wage contracts.

Although the precise terms of the tied-labor contracts observed by Hüsken in Java and Anan in Thailand differ, the logic is essentially similar. In the first place, these arrangements embody built-in mechanisms of labor management that operate according to a principle of exclusion (Hart 1986b). Workers who are incorporated into these arrangements are in a somewhat more secure position than those who depend on the market for casual agricultural labor. Although for brief periods in the peak season wages paid to casual laborers may be relatively high, tied workers have secure employment and, in general, a higher annual income. The existence of a group of people in an economically inferior position is essential to the effective functioning of these arrangements. Confronted with the option of joining the ranks of the excluded, tied laborers work hard to ensure renewal of their contracts. Accordingly, by extending job security on a selective basis, employers ensure themselves of a reliable and hardworking labor force that requires little if any direct supervision. Particularly when employers are actively engaged in nonagricultural activities, this is clearly an important benefit. Another benefit yielded by some types of contracts is that they provide the employer with access to the labor of other household members, most often women and children.

The logic of exclusion and relative privilege undergirds the diffuse patronage relations widely regarded as "traditional" or "precapitalist," as well as explicitly contractualized arrangements. Both Hüsken and Anan document ways in which contracts are being adapted and formalized so as to allow the employer (1) greater control over the production process as nonlabor inputs become increasingly important and (2) appropriation of a larger share of incremental output. The underlying principle of exclusion remains essentially unchanged, however, despite changes in contractual terms. As I have argued more fully elsewhere, this principle also holds in both slack labor market conditions and seasonally tight ones (Hart 1986c).

Exclusionary labor arrangements embody mechanisms not only of labor management but also social control. By creating divisions within the workforce, employers often seek to ensure not only a hardworking labor force but also a loyal and docile one (Hart 1986b). At the same time, however, the process of exclusion is likely to create or exacerbate social conflict and tensions. The state's position vis-à-vis different agrarian classes—as it relates both to the mobilization of political support and to the exercise of "law and order"—is therefore central to explaining the forms of agrarian labor arrangements and the processes through which they change, as well as to interpreting the consequences.

The evidence provided by Hüsken and Anan helps illustrate these points and, in the process, sheds light on some of the contradictions inherent in the agrarian strategies of the Thai and Indonesian states. Anan describes how tenant rebellions that had been widespread in northern Thailand were crushed by the state in 1976. Antagonism and resentment of tenant-cum-permanent workers towards their landlord-employers remain, although in a more muted and suppressed form. Casual workers, however, have been engaging in increasingly overt forms of collective organization and resistance.

The highly differentiated structure of labor arrangements in the village studied by Anan is both a reflection of struggles and antagonisms between workers and employers and a source of new tensions. First, the main modifications of sharecropping arrangements took place in the wake of a series of tenant evictions following the crushing of the tenant rebellion by the state in 1976. The emasculation of tenants' bargaining power attendant upon state intervention on landlords' behalf enabled landlord-employers to redesign contracts so as to appropriate a larger share of incremental output in the main rice season. At the same time, Anan shows how landlords try to maintain an aura of patronly generosity by providing these tenant-workers with off-season concessions that, in some instances at least, elicit outward signs of gratitude and moral indebtedness. The attached-labor arrangements described by Anan constitute an even more overt effort on the part of employers to cultivate loyalty and submission by dividing the workforce.

The increasingly strident demands of casual workers can be seen both as a cause and a consequence of divisions within the workforce. To the extent that exclusionary mechanisms are successful in creating a group of comparatively privileged workers, they may tend simultaneously to intensify resentment and resistance on the part of the excluded. This in turn heightens employers' needs for strategies of co-optation while at the same time rendering these strategies more costly and less effective. Efforts to cut costs by trimming privileges violate the logic of these arrangements and confront employers with a dilemma between accumulation in the short run and ensuring the long-run conditions of accumulation. A process more or less along these lines seems to be unfolding in the village studied by Anan, where the crushing of tenant rebellion by the state has helped create conditions in which new forms of rural resistance are taking shape.

A contradictory set of forces seems also to be emerging in Java, but in a somewhat different form. In comparing the agrarian labor relations observed in Java with those in Thailand, the more openly repressive system of state control in Indonesia is a key consideration. Since the accession to power of the New Order regime in the late 1960's, any form of rural resistance is promptly crushed by the bureaucracy and military, whose coercive and repressive power extends deep into the village. This greater degree of state repression in Indonesia is particularly significant in view of the much higher levels of landlessness in Java which should in principle make exclusionary labor arrangements more difficult to sustain over a long period.

The apparent resurgence of exclusionary labor arrangements in Java in the late 1960's needs to be understood in this context. As I have argued more fully elsewhere (Hart 1986a, 1986c), the mobilization and organization of the poor peasantry by the PKI (the Indonesian Communist Party) in the early 1960's created conditions in which exclusionary arrangements were by and large rendered infeasible. The militarization of the bureaucracy by the New Order, along with the closing down of most forms of local-level political activity and the murder or imprisonment of those accused of communist proclivities, profoundly reshaped agrarian relations and the exercise of power in Javanese rural society. The apparently quite rapid spread of exclusionary arrangements in the late 1960's was, I suggest, primarily a reflection of state repression of organized resistance.

As with their Thai counterparts, the Javanese rural elites' need for a well-disciplined labor force that does not require direct supervision has been intensified by the creation of new opportunities for accumulation, many of them via state patronage. Even though the more concentrated power structure of the Indonesian state places rural elites in a more insecure position vis-à-vis the state, Javanese elites can rely on the coercive capacity of the state to contain organized rural resistance to a far greater extent than those in Thailand. In terms of their relations with less powerful groups, Javanese elites thus appear better able to sustain exclusionary labor arrangements.

Although on the surface the Indonesian state's strategy seems to have been more successful in suppressing rural unrest, it is likely to generate other problems. On the one hand, the suppression of open antagonisms between rich and poor within rural society reinforces the state's need to exercise direct control, which, as we have seen, has partly taken the form of incorporating rural elites into the state machinery as minor functionaries. On the other hand, this process of incorporation could itself generate considerable resentment on the part of dominant rural groups that will become increasingly difficult for the state to alleviate through offsetting benefits. Declining oil revenues have curtailed the state's fiscal capacity to extend patronage and, to the extent that dominant rural groups have developed bases of accumulation independent of the state, the costs of state patronage are likely to be increasing.

State Patronage in Malaysia and the Philippines: The Persistence of Petty Producers

Malaysia and the Philippines also represent a study in contrasts in terms of patterns of subsidy and taxation of the rice sector. Malaysia has the most heavily subsidized rice sector in Southeast Asia, whereas the Philippines has tended increasingly in the direction of extraction. These in turn reflect wide differences in macroeconomic conditions.

In trying to deal with rural dissent, the Malaysian and Philippine governments have been more concerned than their Indonesian and Thai counterparts with building a base of peasant support in the countryside, but they have gone about doing so in very different ways. As discussed by Fegan in the Philippines country introduction, land reform and the imposition of martial law in 1972 were initially designed both to quell peasant unrest and to break the power of the landed oligarchy. By destroying traditional political patronage links and replacing them with peasant cooperatives, Marcos initially sought to develop a new base of support. The Malaysian state in contrast has avoided any direct restructuring of agrarian power relations and has sought instead to build on factional divisions within the peasantry in attempting to undermine orthodox Islamic opposition.

Far from being relatively egalitarian, however, these "peasant-based" strategies have gone hand in hand with the systematic extension of privileges by the state to dominant groups. As in Thailand and Indonesia, the Malaysian and Philippine states have made extensive use of resources associated with the Green Revolution in order to secure the compliance of strategically placed elites.

The character and position of these elites are, however, entirely different. In the Philippines, Marcos underestimated the extent of landownership among the new constituency of bureaucrats, army officers, and professionals that he hoped to nurture alongside the new base of peasant support (Fegan, Chapter 6). To compensate them for land reform, Marcos provided these local elites—based mainly in small towns—with preferential access to a variety of commercial opportunities, many of which were linked with the spread of new rice technology. Small-town elites thus became heavily involved in banking operations associated with huge credit programs, as well as dealerships, marketing, and processing operations stimulated by the state's concerted effort to increase rice production. They are also extensively engaged in a variety of nonagricultural forms of accumulation, including transport, education, trade, construction, and gambling.

Although the Philippines "rural elite" of the 1970's and 1980's were mainly a town-level bourgeoisie, the Marcos state also tried to co-opt effective village political leaders. In addition to channeling some Green Revolution rewards through them, the Marcos state provided these village-level clients with the official connections—to settle disputes, dispense jobs on public works, etc.—that in turn reinforced their position in village society. State policy towards rice producers has been highly contradictory, however. On the one hand the Marcos state hesitated to force land reform beneficiaries to pay their debts, while on the other it moved increasingly in the direction of turning the terms of trade against rice producers.

On the face of it, the Malaysian counterparts of the Filipino small-town elites are Chinese merchants. Far from having been provided with preferential access to commercial opportunities, these merchants have been partially displaced by the spread of state-run operations in input distribution and marketing (Scott 1985). In contrast to the Philippines, rice producers—and particularly dominant groups among them—have been the unambiguous beneficiaries of state patronage in Malaysia.

Although all producers in state-sponsored irrigation schemes like the Muda area discussed by Lim and Ikmal have benefited from subsidized water, fertilizer, and rice price supports, a disproportionate share of state resources have been channeled to leading representatives of the state party UMNO (United Malays National Organization) in the countryside. This in turn is part of the state's strategy to develop an indigenous bourgeoisie, which in turn needs to be understood in terms of the particular ways in which ethnic and class forces combine at the national level in Malaysia.

Perhaps even more importantly, the channeling of resources to the UMNO elite is designed to undermine the orthodox Islamic party (PAS or Partai Islam). As noted by Scott (1985), a key presumption of the Malaysian state strategy has been that the provision of resources to the upper strata of rural society is sufficient to ensure that they will mobilize the support of the small peasantry.

The case studies, along with a recent study by Scott (1985), help elucidate the links between state-sponsored accumulation by rural elites and the conditions of existence of small producers in the main rice-growing regions of Malaysia and the Philippines. In so doing, they also shed light on some of the ways state patronage tends to undermine itself through the generation of unintended consequences.

In both countries, the late 1970's represent an important turning point. Until then, state patronage structured around the dissemination of modern rice technology was reasonably successful in ensuring both rice production and a base of peasant support in the countryside. Since then, however, the agrarian strategies of both the Malaysian and Philippine states have been coming apart, but in very different ways.

On the face of it, the holding capacity of the small rice peasantry in Malaysia and the Philippines during much of the 1970's was simply a reflection of support by states intent on ensuring a docile constituency in the countryside. Land reform in the Philippines provided a segment of former tenants with relatively secure property rights and in some instances the power to resist paying rents and even to engage in land takeovers (Banzon-Bautista, Chapter 7). By the same token, large irrigation schemes in Malaysia enabled small producers to resist dispossession through indebtedness by significantly increasing productivity (Lim, Chapter 9).

At the same time, however, we have seen how the strategies pursued by the Malaysian and Philippines governments were "peasant-based" only in a very limited sense. Both governments have been vitally concerned with securing the support of strategically placed groups, and a considerable proportion of state resources allocated to the rice sector has been channeled through these groups.

In the case of the Muda Irrigation Scheme in Malaysia, Lim notes that discrimination in access to subsidized fertilizer operates along party political lines, rather than against small producers. Several studies reveal similar patterns in the distribution of other resources disbursed by the state (Scott 1985; Wong 1983). Several sets of forces seem to have been particularly important in ensuring that at least a segment of the small peasantry gained access to state-subsidized resources. First, tightening of the labor market—as a consequence both of intensified demand for agricultural labor and of industrial growth that has drawn off a sizable proportion of the rural population—enhanced the relative bargaining position of the small peasantry until the late 1970's. Second, in order to maintain positions of power at the local level, dominant rural groups have had to ensure a base of political support. Scott (1985) shows how, until the late 1970's, dominant rural groups actively cultivated their poorer neighbors, both in order to ensure an adequate labor force and to consolidate their political support.

An entirely different set of mechanisms provided small peasants in the Philippines with access to state-subsidized resources. In contrast to Malaysia, where the position of dominant rural groups is in part contingent on their ensuring a political following from within rural society, political patronage links in the Philippines were sharply reduced by martial law. What does seem to have been important in ensuring that small peasants gained access to subsidized credit prior to 1979 was both the sheer size of the program as well as the way in which credit was distributed. As Fegan explains, extending control over the banking system was one of the important ways in which Marcos sought to placate members of the former landowning class in the wake of land reform. The Masagana-99 program not only resulted in a huge expansion of banking business; it was also organized in such a way that default was extremely profitable and the indiscriminate extension of credit was very much in the interests of those who controlled the banking system. Wurfel (1985) has noted that the profits derived by local elites from banking and other commercial operations far exceeded the interest subsidy to rice cultivators.

By the late 1970's, the mechanisms through which at least some of the small peasantry gained access to subsidized state resources had helped create conditions that both undermined these mechanisms and opened up new opportunities for accumulation by rural elites. The ways in which rural elites have sought to take advantage of these opportunities have in turn undermined the agrarian strategies of both states.

In the Philippines, the massive misallocation of resources associated with the Masagana-99 credit program contributed to the deteriorating macroeconomic conditions which rendered the program increasingly infeasible. By the end of the 1970's subsidized credit had shrunk considerably, at the same time as the terms of trade were increasingly moving against rice production (Fegan, Chapter 6).

In the main rice-producing area of the Philippines, Fegan shows how those with an investable surplus are shunning direct production as increasingly unfavorable price ratios have reduced profitability. Those who continue to invest in agriculture are taking advantage of opportunities for accumulation that have been greatly enhanced by the drying-up of most institutional credit. Although the breakdown of the institutional credit system generated enormous demand from small producers for alternative sources of credit to purchase fertilizer and chemicals, the problem was collateral. Lenders were unwilling to accept usufruct rights in land as collateral not only because of the declining profitability of rice production but also because of the uncertainty surrounding property rights in land. This problem was resolved by the mobile threshing machines that allow lenders to collect at harvest time. Fegan's paper contains detailed evidence on the ways in which rural capitalists are accumulating across linked markets in credit, inputs, machine services, and trading—with the help of an infallible collection mechanism—while simultaneously standing back from direct production.

Fegan also shows how landholdings have become more dispersed in the village in which he worked; with the rice crop rather than land as collateral, small peasants are less likely to lose control over land. While formally still in control of the means of production, small peasants seem caught up in a process closely analogous to dispossession.

The mechanisms through which local elites intensified their extraction from the peasantry have both been shaped by and contributed to the process of rural economic decline. This process assumes particular significance when viewed in the context of the fiscal crisis of the Marcos regime.

The undermining of the Malaysian state's agrarian strategy has come about in an entirely different way. In the Muda region of Malaysia we have seen how, prior to the late 1970's, the rural elites' need to ensure an adequate labor supply led them to reinforce "traditional" patronage relations with small peasants (Scott 1985). From the viewpoint of state strategy, this elaboration of patronage relations within rural society was a distinct advantage to the extent that it enhanced the capacity of dominant rural groups to bring the poorer peasantry along politically. At the same time, however, the enormous intensification of labor needs and rising labor costs brought about by the combination of irrigation, double-cropping, and modern rice technology placed limits on the ability of rural elites to accumulate within agriculture.

The introduction of combine harvesters in the late 1970's, actively encouraged by the state agency in charge of administering the Muda scheme (MADA or Muda Agricultural Development Authority), has drastically reduced labor needs for harvesting and threshing. MADA is also encouraging the development of broadcast sowing technology, thus doing away with highly labor-intensive transplanting of rice seedlings. Most of the labor displaced by mechanization has been that of older women.

Mechanization has not only enabled large landowners to expand their scale of operations (Ikmal, Chapter 11) but has also resulted in the dramatic restructuring of agrarian relations. Scott describes how rural elites are trying to disengage themselves from patronage relations with the small peasantry now that the latter's labor has become increasingly irrelevant. Although UMNO leaders still need to maintain a base of political support within rural society, they are trying to limit the necessary obligations through what Scott terms a minimum winning coalition strategy. In the process, their capacity to bring the poorer peasantry along politically has sharply declined.

For their part, small peasants are actively trying to invoke older patronage obligations with very little success, which in turn is giving rise to considerable resentment. Although open rebellion seems unlikely, the forces at work within rural society appear increasingly problematic from the viewpoint of the state.

In the first place, small peasants may refuse to conveniently disappear, as advocates of mechanization within the state agency expected that they would (see, for example, Afifuddin 1974). According to these advocates, small peasants displaced by mechanization would simply be absorbed into nonagricultural activities stimulated by agricultural progress. Contrary to these expectations, relatively little diversification has taken place in the Muda economy and there have instead been large capital outflows (Bell et. al. 1982). Although research into the fate of the small peasantry in the wake of mechanization is still quite limited, there are indications that many of them may resist full proletarianization through a strategy of part-time farming. In particular, what seems to be happening is that men and younger women from poor rural households are migrating on a temporary basis to industrial and construction jobs in urban areas, leaving older women in charge of farming.

An important pattern suggested in the evidence presented by Lim and Ikmal is that small peasants hire combines and tractors and that machinery rental contracts are sometimes tied in with credit arrangements. Although some large Malay landowners have diversified into renting out machinery, the market is dominated by Chinese. The diffuse patronage-cum-labor relations within the rural Malay community that were reinforced in the first phase of state patronage have thus been replaced by a set of relations that, from the viewpoint of the state as it is now politically constituted, are not only less reliable in terms of ensuring UMNO support but also potentially more volatile.

The UMNO elite in the countryside are not only less capable of exercising social and political control within rural society, but they have failed to develop into the independent bourgeoisie envisaged by the New Economic Policy. Instead, according to Scott, they have by and large remained heavily dependent on state support and are likely to make strident demands on the state for continued support. At the same time, declining oil revenues are curtailing the state's capacity to subsidize rice production.

These growing fiscal constraints, along with the declining growth of industrial employment, are particularly significant in view of the existence of a counterelite. Dominant PAS groups have benefited from certain forms of state patronage, but they remain opposed to the ruling coalition. Whether the PAS leadership moves to mobilize the support of the poor peasantry—and the way in which the ruling coalition deals with such a challenge—are likely to exercise a critical influence on future processes of agrarian differentiation.

Conclusions

Comparative study of different Southeast Asian cases has shown how agrarian change can follow a number of different paths, depending on the configuration of external political-economic forces and on internally generated processes. In order to interpret these divergent trajectories, we need to specify how forces operating at different levels are connected with one another. In particular, we need to understand how external forces not only influence the forms of institutional arrangements within rural society but are in turn affected by them.

In the Southeast Asian context, the multiple and changing forms of state patronage contain the key to identifying and interpreting these connections. By drawing on evidence from the case studies, I have tried to show how the particular forms of state patronage (which are themselves a reflection of power struggles in the past) influence mechanisms of extraction and accumulation, and how these in turn generate contradictory and unintended consequences. Although this nonlinear dynamic emerges in all four cases, the particular forms of state patronage and their consequences vary according to historically specific political and economic conditions. Thus, for example, state-sponsored accumulation by dominant rural groups has undermined their ability to bring the rest of the peasants along politically (Malaysia), reinforced agrarian backwardness (the Philippines), and created new but different problems of social control (Thailand and Java).

The comparative material also shows how very different mechanisms can produce a superficially similar phenomenon. The persistence of petty producers in major rice-growing regions in Malaysia and the Philippines is a case in point. We have seen how petty producers can persist in widely different economic and political circumstances, and for reasons that change over time within a given setting. The wide variations in the conditions of persistence that the evidence reveals, together with the dynamism of these conditions and the institutional arrangements associated with them, underscore the limitations of the numerous efforts to develop a general theory of the persistence of petty production.[4]

In some instances it is possible to identify important similarities in the operation of particular mechanisms in different circumstances. For example, the logic underlying the exclusionary labor arrangements observed in the Thai and Javanese cases is essentially similar. Understanding the internal logic of a particular arrangement does not, however, provide a basis for predicting or deducing its long-term consequences for rural differentiation and class formation. Exclusionary labor arrangements in the Thai village studied by Anan have given rise to open expressions of dissent and solidarity among those excluded from these arrangements, while the highly centralized structure of state power in Indonesia precludes any such overt resistance and organization by subordinate classes.

As suggested earlier, however, the suppression of resentment within rural society is likely to become increasingly problematic for the Indonesian state, particularly as oil revenues decline.

Taken as a whole, the comparative Southeast Asian evidence suggests very strongly that we relinquish the search for a single, deterministic theory of agrarian change and differentiation and focus instead on the conceptual and methodological issues and problems involved in trying to understand local-level processes in the context of the wider political and economic forces which they both reflect and alter. It is within this context of viewing change as a dialectical rather than a linear process that explicit attention to the state's role in rural differentiation is so crucial. Bringing the state into the analysis is not simply a matter of viewing agrarian processes as "an epiphenomenon of state power" (Brass 1983). Nor is it a question of the "inefficiency" of state intervention (Bates 1981). Rather, it entails understanding how power struggles at different levels of society are connected with one another and related to access to and control over resources and people.

Notes

1. The forms of rural dissent in the four countries are discussed more fully in the country introductions.

2. In this paper I am focusing particularly on the period since the late 1960's, although some form of patronage by the state of dominant rural groups is much older than this.

3. In this volume, see also White and Wiradi (Chapter 13) and Banzon-Bautista (Chapter 7). Other examples are documented in Hart (1986b).

4. For a review of efforts along these lines, see Goodman and Redclift (1982).

References

Afifuddin b. Haji Omar (1974). "Some Implications of Farm Mechanization in the Muda Scheme." In M. Barnett and H. Southworth, eds., *Experience in Farm Mechanization in Southeast Asia.* New York: Agricultural Development Council.

Anderson, B. (1983). "Old State, New Society: Indonesia's New Order in Historical Perspective." *Journal of Asian Studies* 42:477–496.

Bates, R. (1981). *Markets and States in Tropical Africa.* Berkeley and Los Angeles: University of California Press.

Bell, C., et al. (1982). *Project Evaluation in Regional Perspective.* Baltimore: Johns Hopkins Press.

Brass, T. (1983). "Permanent Transition or Permanent Revolution: Peasants, Proletarians and Politics." *Journal of Peasant Studies* 11:108–112.

Girling, J. (1985). "The Political Economy of Rural Thailand." Paper presented at SSRC conference on National Policies Towards the Agrarian Sector, Thailand, January 1985.

Goodman, D., and M. Redclift (1982). *From Peasant to Proletarian: Capitalist Development and Agrarian Transitions.* Oxford: Basil Blackwell.

Hart, G. (1986a). *Power, Labor, and Livelihood: Processes of Change in Rural Java.* Berkeley: University of California Press.

——— (1986b). "Interlocking Transactions: Obstacles, Precursors or Instruments of Agrarian Capitalism?" *Journal of Development Economics* 23:177–203.

——— (1986c). "Exclusionary Labour Arrangements: Interpreting Evidence on Employment Trends in Rural Java." *Journal of Development Studies* 22:681–696.

Robison, R. (1978). "Toward a Class Analysis of the Indonesian Military Bureaucratic State." *Indonesia* 25:17–39.

Scott, J. (1985). *Weapons of the Weak: Everyday Forms of Peasant Resistance.* New Haven: Yale University Press.

Wong, D. (1983). "The Social Organization of Peasant Production: A Village in Kedah." Unpublished Ph.D. thesis, University of Bielefeld.

Wurfel, D. (1985). "The Political Economy of Philippine Agricultural Policy." Paper presented at SSRC conference on National Policies Towards the Agrarian Sector, Thailand, January 1985.

Part Two

Thailand

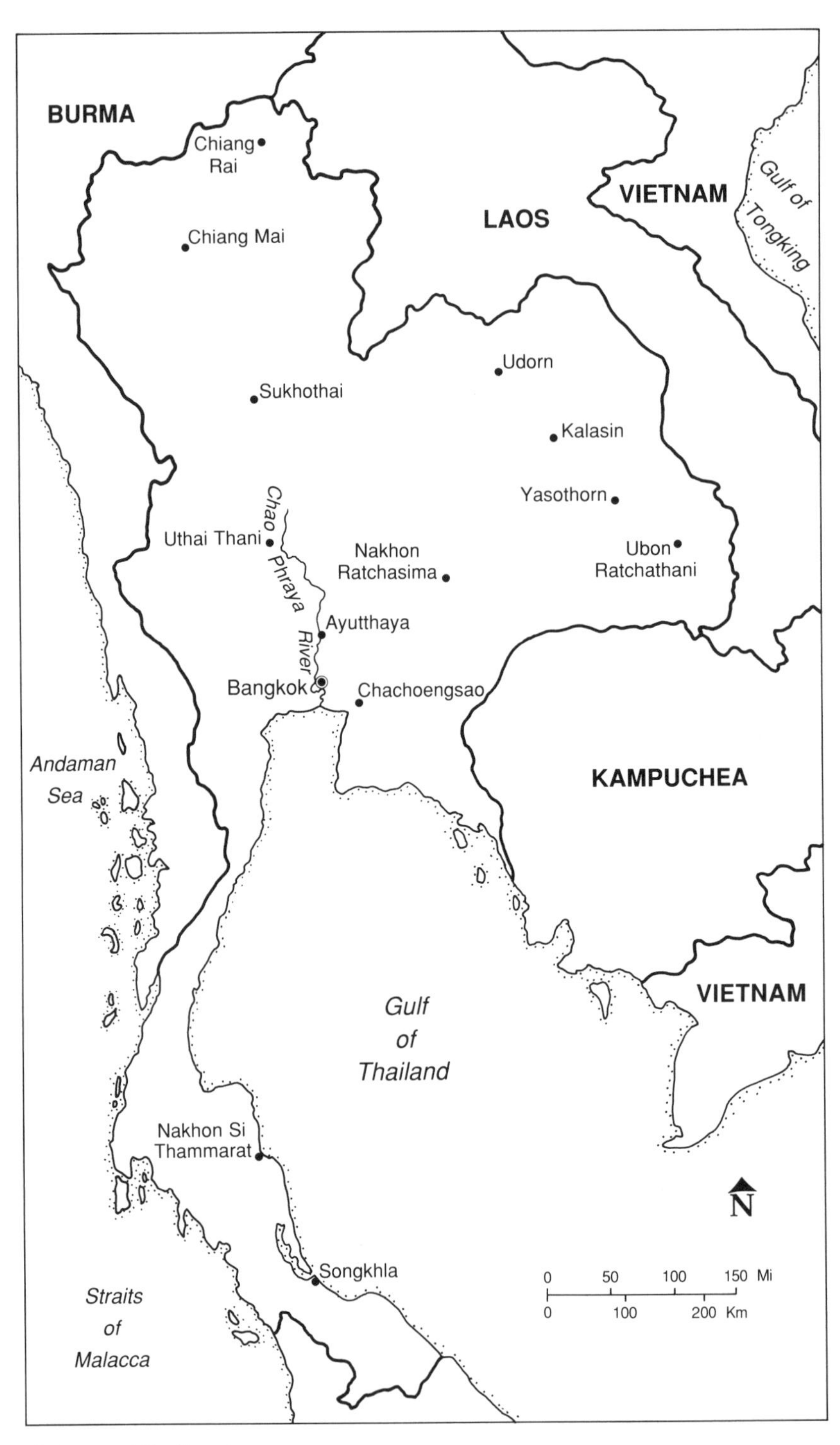

Map 1. Thailand

Chapter Three

Thailand: Agrarian Bases of State Power

Andrew Turton

For centuries rice farming has been the major productive activity and way of life of the majority of Thai people. In the mid-1980s more than half of Thailand's working population are still engaged in agriculture, most of them producing rice as a main or subsidiary crop, and rice remains the most important single crop by value (World Bank 1980:8). In 1982 Thailand was the world's fifth largest producer of rice after China, India, Indonesia, and Bangladesh and is one of the highest net agricultural exporters among developing countries—in 1978 the highest in Asia (Hawes 1982). At least one-third of agricultural production is exported (Douglass 1984). Yet in the past twenty-five years the relative position of agriculture within the national economy, and that of rice within the agricultural sector, have undergone profound changes.[1]

Some distinctive features of the rice sector may be highlighted at the outset. First, despite the volume of the marketed and exported rice surplus, productivity is one of the lowest in the world, as is the adoption of modern technology in rice production. Second, the agricultural sector, and perhaps especially rice within it, has been massively neglected by governments, at least in the positive sense of subsidies, supports, and reforms directly aimed equitably to benefit producers. Third, despite a high degree of commoditization of the rural economy, and against the expectations of some classical theory, there is a preponderance of owner-operated smallholdings, although these are relatively large compared

TABLE 3.1. *Distribution of Agricultural Households: 1978*

Region	No. of agricultural households	Percent agricultural households	Percent land area	Percent agricultural land
Center	875,449	20.29	20.2	25.1
North	1,124,100	26.06	32.9	20.8
Northeast	1,724,558	40.01	33.1	42.4
South	588,184	13.64	13.8	11.7
Total	4,313,291	100%	100%	100%

Source: After Witayakorn (1982:4, 92).

with many other rice-producing countries. Fourth, there has been a historical reluctance for big capital to be invested in direct production of rice excepting some infrastructural development (notably irrigation)—though for the first half of the twentieth century the rice industry as a whole (rice growing, milling, exporting) was the most important industry and contributed most to the formation of indigenous capital (Suehiro 1985).

Agriculture has declined in relative importance within the national economy since the early 1950s when it accounted for just over 50% of GDP; by 1982 it had fallen to 20%, compared with 27.7% for industry (Far Eastern Economic Review 1984). Agricultural growth is expected to continue at its historical rate (4–5%) during the 1980s (Panayotou 1985), though at a rate considerably lower than the industrial and service sectors.

In the early 1950s, 88% of the labor force was engaged in agriculture; by 1982 the proportion had fallen to 71.6% and is projected to continue to fall. Nevertheless agriculture remains of strategic importance. Over 80% of the population still live in rural villages, the great majority engaged in agricultural production, and the absolute size of the rural/agricultural population is increasing at somewhat under 2% annually. Two-thirds of agricultural households are in the northern and north-eastern regions with which the two papers that follow are primarily concerned. Table 3.1 shows the regional distribution in 1978 of population, agricultural land, and households, providing a base against which other regional distribution figures given below can be assessed.

One of the tasks of this introductory chapter and the two following papers is to show how aggregate statistics, rates, projections, etc., conceal disparities, inequalities, and forms of differentiation between regions, among producers, and between producers (whether owners of land or not) and those who control strategic means of production, means of surplus extraction, and the means of social and labor discipline.

The Rice Sector

Rice production has declined in relative importance within the agricultural sector. The share of rice among all exports, which up to 1950 had been 50% or more, declined to 20% in 1969, while its average share among other crops declined from 49.1% in 1960–1965 to 36.7% in 1976–1979 (World Bank 1980:8). Although the area planted to rice has risen in absolute terms, it has declined relative to other crops (Ingram 1971; World Bank 1983; Thailand 1984). This diversification has been assisted by government support for export crops (price supports, encouragement of agrobusiness, etc.) and high world market prices in the early 1970s.

For almost a century after the Bowring Treaty of 1855 between Thailand and Britain—the first of many trade treaties with industrial countries—central Thailand had been the principal rice-growing region. The region accounted for most of the estimated 25-fold increase in the volume of paddy production from 1850 to 1934 and the near quadrupling of the area planted to rice (Ingram 1971). Until 1950 the almost monocultural central region produced considerably more paddy than all other regions. The northern and northeastern regions developed rice production for the market later—due to limited irrigation and transport facilities among other factors—from more diversified and predominantly subsistence production, and while now diversifying in turn into cash and export crops these regions have also come to be the main rice-growing areas. In 1983–1984 central Thailand accounted for only 22% of the main-season rice production but nearly 80% of the second crop because of superior irrigation systems.

Despite diversification and relative decline of rice compared with other crops, the area planted to paddy increased from 5.75 million ha in 1958–1959 to 9.30 million ha in 1983–1984, and volume of paddy production grew from 6900 tonnes to 16,900 tonnes in the same period (Thailand 1984). The 1978 Agricultural Census shows 81.93% of all agricultural holdings planting some rice—47.77% as a single crop, 34.16% com-

TABLE 3.2. *Regional Patterns of Rice Production: Area and Number of Holdings Planted to Rice by Region, 1978*

Region	No. of holdings planted to rice only, as % all holdings in region	No. of holdings planted to rice only and rice with other field crops, as % all holdings in region	Total area planted to rice, as % all agricultural land in region
Northeast	54.59	91.65	73.5
North	50.58	83.23	66.5
Center	45.45	64.91	58.3
South	25.95	74.38	37.0

Source: Thailand (1978).

bined with other field crops. The extent to which agricultural households in different regions are involved in rice growing can be judged from Table 3.2, which clearly shows the continued importance of rice as both cash and subsistence crop for the great majority of farmers—especially in the North and Northeast—even when they have diversified into other major cash crops. Approximately one-third of the total paddy area is planted to glutinous rice, almost all of which is grown in the Northeast (where it amounts to over half of all rice produced) and in the North (approximately one-third). Approximately 80% of all paddy (and nearly all glutinous rice) is transplanted. Upland rice accounts for less than 2% of the total paddy area.

Aggregate figures show Thailand to have one of the lowest rates of productivity (area yield) in the world, comparable with Brazil, Burma, and India—about one-third that of Japan, the two Koreas, and the U.S.; half that of China; and considerably lower than Vietnam or Indonesia. Table 3.3 shows considerable variations in yield in the different regions; yields in rain-fed areas are lower still. It could be said that Thailand has scarcely undergone even a modest Green Revolution. In 1976, high-yielding varieties (HYVs) had been adopted on less than 5% of grain-producing land; and in 1983–1984 only 7.7% of paddy land was double-cropped with rice—75% of it in the central region—producing 12.8% of total paddy production. Thailand has one of the lowest levels of irrigation in Asia. More than 60% of paddy land is not irrigated, and

TABLE 3.3. *Regional Patterns of Rice Production: Rice Area and Yields by Region, 1984*

Region	Area planted (%)		Production (%)		Yield (kg/ha)	
	Main crop	*Second crop*	*Main crop*	*Second crop*	*Main crop*	*Second crop*
Northeast	52	8.6	43.7	5.7	1562.5	2468.7
North	22	12.5	28.7	12.4	2443.7	3712.5
Center	20	74.7	22.2	78.9	2300.0	3893.7
South	6	4.2	5.3	2.9	1687.5	2593.7
Whole country	100	100	100	100	1906.2	3693.7

Source: Thailand (1984).

the regional shares are skewed—for example, with the Center (25% of all agricultural land) having 68% and the Northeast (over 40% of all agricultural land) having less than 10% of irrigated area (Panayotou 1985, citing Royal Irrigation Department).

Use of fertilizer—an average of only 17 kg/ha—is also lower than in almost all Asian countries. Fertilizer prices are high and most fertilizer is imported. In 1983 about 58% of all fertilizer used was in paddy production, of which 30% (18% of the total) was for the second crop. Again the regional distribution of fertilizer use shows the now familiar imbalance, with most fertilizer being used in irrigated areas. Because of the high cost of fertilizer (made worse by the still relatively low availability of institutional credit and high cost of noninstitutional credit) and poor quality, the use of fertilizer (and other inputs) is often below optimal levels.

Mechanization of agricultural production—primarily tractorization—is increasing rapidly in the Center, but draft animals are still widely used in the North and Northeast.

Land and Labor

Problems of analyzing the significance of different kinds and degrees of landownership, tenancy, and landlessness are addressed in detail in the papers which follow; here only some broad trends will be indicated. Broadly speaking, Thai agriculture has historically been predominantly characterized by self-sufficient family farms with customary and sometimes legal notions of maximum holdings of 4 ha (in some periods and regions 8 ha)—with perhaps 1 ha, in some regions at least, affording a viable subsistence minimum. Much of the increase in production has been accounted for by an increasing rural population expanding into and developing new land for agriculture, and this extensive agriculture continues to be characteristic.

Although there has not been a marked overall trend towards a polarization between large landowners and a correspondingly large class of landless agricultural laborers, there is nonetheless an unequal distribution of land. The last agricultural census in 1978 showed that nationally 43% of farmers operate only 13% of land, while the largest 16% operate 44% of land. Census figures are for operated farms and do not indicate the extent of concentration of agricultural landownership. Between the 1963 and 1978 censuses there had been increases in average farm size and in the number of medium and large farms and a decrease in the number of small farms. An optimistic reading of this might be that small farms had grown in size and become more economical units. Many small farms have disappeared altogether, however, and the rate of farm household formation is well below that of the increase in rural population. This is true even for the North and Northeast, which between 1963 and 1978 accounted for 90% of the increase in farmed area and 80% of the increase in farmholdings. Moreover smaller farmers are likely to have less legal security on land owned, and indeed they are likely not to own all or part of the farms they operate. The quality and accessibility of their land is likely to be worse. They are likely to have less access to irrigation, cheap credit, and inputs, and so to be restricted to single-cropping with lower yields. To these factors we must add the regional inequalities in access to irrigation and modern inputs noted earlier.

While there is not a high incidence of large plantations or very large single farms, there are still a few hundred old landed families who own estates of areas totaling around 500 ha or more, often acquired through seizure or royal grant in the latter part of the nineteenth century. And there is a new phenomenon of estates of comparable size being purchased by agrobusiness companies and other new urban-based owners. The land situation has to be seen in the context of the near exhaustion of land reserves available or suitable for agriculture, the existence of which has hitherto mitigated many basic agrarian problems.

Family labor constitutes the greatest proportion of farm labor. The 1978 Agricultural Census shows only 29% of farms reporting the use of any hired agricultural labor. Local studies indicate a considerable increase in the use of wage labor in agriculture since 1978, a continuing decline in the use of exchange labor, and many wage-labor-employing households also providing wage labor in turn.

The census undoubtedly underreports the extent of virtually landless or extremely land-poor households, since it excludes a number of small-scale or subsistence operations under 0.32 ha where annual income earned from *selling* farm products was less than 5000 baht; it also excludes 'hilltribe villages located on hills' and whole 'villages whose number of agricultural holdings was less than 25% and such percentage was less than 25' (Thailand 1978, Whole Kingdom volume, 'Coverage of the Census'). Estimates of completely landless rural *households* vary between 10 and 25% or more. The World Bank (1978) reports that in 1975–1976 some 17% of all village households had no agricultural enterprise at all, though this figure would include households of shopkeepers, officials, and teachers and other nonpoor households and exclude landless *agricultural* laborers. The same report indicates that some 40% of the incomes of households engaged in agriculture was from nonfarm sources; this is over 50% for the Northeast and South. Such figures are for households, while many individual members of households owning some land are effectively landless or are part-time agricultural workers. Moreover a large proportion of the unpaid family workforce of landowning households are virtually unemployed, on the farm or off, for at least three months in a year and often much longer—from under one-third of the workforce in the Center to two-thirds in the Northeast in 1979 (World Bank 1983:table A13). These figures do not feature in national unemployment statistics; nor do they include members absent for long periods as migrant workers (or in search of work) who may return to the countryside. The latter include several hundred thousand young women engaged in various forms of prostitution (see Pasuch 1982; Omvedt 1986).

Figures for agricultural tenancy are also problematic. Table 3.4 shows that in 1976 some 20% of all agricultural households may have been tenants or part-tenants, and in the same year just under 12% of total land in holdings was reported to be operated by tenants. Several provinces in the Center have a tenancy rate of over two-thirds, while the poor Northeast has a low tenancy rate of less than 10%. The figures

TABLE 3.4. *Proportion of Tenants and Part-Tenants by Region*

Region	1973 (%)	1976 (%)
Center	29.27	41.31
North	15.50	26.71
Northeast	3.27	8.68
South	4.42	17.68
Whole country	12.27	20.84

Sources: Thailand (1973, 1976).

are higher in certain paddy-growing areas in the North (e.g., Chiang Mai) and the Center (e.g., Ayutthaya) where in certain districts the overwhelming majority, and in some villages all operators, may be tenants, sometimes of a single owner. The numerous intricacies and variations of forms of tenancy are addressed in the following papers. While some larger tenants, especially part-tenants, may be successful commercial farmers, and some forms of 'filial' tenancy may represent a relatively low-cost mortgage on heritable land, most forms of tenancy are probably disadvantageous for the tenant vis-à-vis the landowner or compared with the nontenant farmer. The Agricultural Land Rent Control Act of 1974, which fixed a maximum rent of one-third of the crop, has been massively ignored by both government and landowners.

Following the 1975 Agricultural Land Reform Act—passed precipitously as a result of farmers' mobilizations in the previous few months—an Agricultural Land Reform Office was established. Little has been achieved since then, however, and it may be doubted whether the political will exists to implement even the most limited of land reform programs on a scale commensurate with the problems of landlessness and tenancy. Potential Land Reform Areas have decreased in size, and there has been a shift in emphasis from the transfer of privately owned land to the utilization of 496,000 ha of encroached Forest Reserve Areas. By 1981 only some 30,000 ha of land in the Reform Areas had been purchased from private owners, comprising about 10% of land in the Reform Areas and under 2% of all rented land, benefiting under 3% of tenant households.

Government Policies Towards Agriculture and the Rice Sector

Following the Bowring Treaty in 1855, the encouragement of paddy production and export became major government policy. Some of the more important elements of early policy included legislation (elimination of slavery and corvée labor by the first decade of the twentieth century), land taxation, and capital expenditure—e.g., on irrigation and to some extent transport (Ingram 1971). These policies contributed to the enormous increase in the area planted to paddy and volume of paddy production and almost monocultural development, in the Center at least, which we have already noted. By keeping the domestic price of rice low and favoring oligopolistic rice exporters, government policies assisted a massive transfer of wealth from the paddy producers: in 1951 rice producers are estimated to have received about half the paddy price they would probably have received in the free market (Ingram 1971).

Up to the turn of the twentieth century, European trading houses monopolized rice milling, export, insurance, and shipping, but this was largely taken over by resident Chinese merchants and formed the basis of the accumulation of merchant capital and the founding of banks in the 1920s and 1930s. This ethnic factor—agricultural landownership was also reserved for Siamese/Thai citizens—contributed to the nonintervention of capital in direct paddy production. Nor did the few wealthy farmers become merchant capitalists (Suehiro 1985). Since the World Bank's development strategy was adopted in the first national development plan in 1961, successive policies have neglected or discriminated against the agricultural sector and favored urban and industrial development, the accumulation of capital outside the rural sector, and the integration of the Thai economy into the world economic system. Now the World Bank itself underlines how discriminatory government policies on agricultural prices, industrial protection, and trade have resulted in a huge net transfer of wealth from the countryside (World Bank 1983).

Historically only some 10% of government investment has been in agriculture. A World Bank report (1983) attempts to quantify the effects of government policies in favor of rural areas, giving approximate orders of magnitude of capital expenditure for an unspecified year (1980?). About half the total is spent on irrigation (the Royal Irrigation Department is one of the largest in the Ministry of Agriculture and Cooperatives), about one-quarter (in 1980) on the off-season Rural Employment

Generation Program, and most of the rest on interest subsidies on loans to agriculture. Institutional credit has expanded considerably in recent years (from 371 million baht in 1967 to 11,000 million baht in 1980), and the Bank of Thailand stipulated in 1978 that 11% (raised from 5% in 1975) of commercial bank deposits should be lent to agriculture. In 1977, 25% of farmers were reached by institutional credit, mainly through the Bank for Agriculture and Agricultural Cooperatives set up in 1966.

Some of the inequitable distributional consequences of government policies have already been seen, whether in credit, irrigation, fertilizer, research and extension, or land policy (land titles, land reform). Various (usually ad hoc) price guarantee systems have reached few producers. In 1982–1983 only 7% of paddy was purchased at supported prices. In the same year only about 10% of fertilizer used was distributed by the Marketing Organization for Farmers. Apart from the gross regional inequities already emphasized, most of the specific government benefits reach the better-off producers. Delivery is especially through government-controlled agricultural cooperatives and Farmers' Groups in which larger, wealthier farmers tend to predominate as members and especially as controlling committee members. In any case only a small proportion (some 15%) of agricultural producers are members of cooperatives, and only some 11% of rice producers are members of Farmers' Groups specializing in paddy production (365,262 members of 2494 groups in 1982).

Theodore Panayotou (1985) has characterized the contradictory, uncoordinated, and short-term nature of government agricultural and especially rice policies. He summarizes the four principal elements of rice policy as: the regulation of rice exports; the maintenance of a supply of low-priced rice for domestic consumption; the maintenance of government revenue from rice export taxes; and at the same time the maintenance or increase of farm-gate paddy prices. Paddy farmers are squeezed on the one hand by the control of rice export prices, which determines a low domestic price, and the export tax on rice (the 'rice premium' initiated in 1955)—which for many years accounted for 20–30% of the export price (though this has recently declined) and which is passed down the line to the farmer by exporters and millers. On the other hand, farmers are disadvantaged by industrial and trade policies which tax agriculture by increasing the price of production inputs and consumer goods.

One of the notable beneficiaries of government policies has been agrobusiness (feed and flour milling, livestock farming, manufacture of canned pineapples, tapioca pellets, etc.), which has been the fastest-growing sector since 1975. Investing in agrobusiness by indigenous capital and transnational corporations—often working with Thai partners as 'joint ventures'—was spectacularly encouraged by the Investment Promotion Act of 1972 (revised 1977), which allows *inter alia* for land-ownership by foreign firms, transfer of profits, and tax exemptions (Pasuch 1980). The structure of the agrobusiness sector is highly oligopolistic, with one Sino-Thai transnational conglomerate (Caroen Phokaphand) controlling 36.4% in 1981 and the top five companies accounting for 63.4% (down from 75% in the previous five years) (Suehiro 1985). These companies have had a marked impact on nonirrigated upland crop production (e.g., cassava, sugarcane), on crops such as pineapple and tobacco, and on the production of animal feed and poultry raising. While sometimes themselves owning large amounts of land, these companies purchase crops from small peasant producers and large commercial farmers, but increasingly the latter (especially for certain crops such as sugar). For the smaller producer the monopolistic control—not only of markets and prices but also of an increasingly large number of strategic factors of production—by these companies tends to lead to growing indebtedness and dependence. It often leads eventually to refusal by the companies to purchase their products at all, as difficulties are encountered with meeting demands for quotas, qualities, delivery dates, etc. In addition, subsistence production is undermined and there is incommensurate expansion of opportunities for wage labor; such opportunities as do exist offer low wages and insecure conditions of employment.

Akira Suehiro has analyzed the process of capital formation in Thailand in a way which allows us to consider the relationship between big capital and agriculture. He characterizes a 'tripod structure' of dominant capital in Thailand in which state, local big capital, and foreign or multinational capital are linked (Table 3.5). Suehiro shows a marked dependence of Thai industrial conglomerates on foreign technology, foreign capital, and imported machinery. He also suggests that the pattern of capital accumulation by Thai (light) 'industrial conglomerates' and 'financial conglomerates' is similar to that of commercial capital—namely by their focus on handling goods rather than on the production process itself; by their market orientation; by their inclination towards

TABLE 3.5. *Largest 100 Firms: 1979*

	Assets		Sales	
	Number of firms	*Value of assets (%)*	*Number of firms*	*Value of sales (%)*
State capital	19	36.3	19	22.8
Local private capital	52	53.6	44	31.9
Foreign/multinational	29	10.1	37	45.3

quick returns; and by their lack of attention to innovative technology. All these features, combined with other policies and factors considered above, would seem to point to a relationship, or relationships, between state, capital, and industry on the one hand and agriculture on the other, in which capital accumulation in agriculture is restricted and surplus extraction from agriculture is maximized. Big capital thereby increases its control of production while avoiding large-scale involvement in direct production.

Moreover, recent development has been to extend cultivation into the dry season rather than to increase yields; and industrial development has failed to provide sufficient employment—especially in rural areas—for growing numbers of rural landless, unemployed, and underemployed people. One relatively new area of employment is the 'export' of labor. In 1983 there were an estimated 300,000 Thai workers in the Middle East alone, who were expected to remit 20 billion baht—approximately the value of rice exports in that year (Far Eastern Economic Review 1984).

Political Aspects of Government Policies Towards Agriculture

While some incremental benefits may have had a fairly wide, if limited, distribution, the net social and economic consequences of policies affecting the agricultural sector—and absence of policies—have been to maintain or create massive regional disparities (notably affecting the Northeast, which contains over one-third of total population and an even greater proportion of the agricultural population). This has led to increased inequalities of income and unequal access to means of production (especially in favored regions where capitalist development in agriculture is more advanced), access to employment, and indeed access to virtually all social benefits, including health and education. Some estimates suggest that there may be 10 million people in rural areas living in absolute poverty, and there is considerable malnutrition and death from malnutrition, especially among rural children. There is also much policy rhetoric about 'economic justice' and the plight of the rural poor, more substantial than hitherto perhaps in the fifth National Economic and Social Development Plan (1982–1986). But an authoritative report on development strategy for the 1980s (Somsakdi et al. 1983:114–115) indicates that "in the coming decade, the strategy for agricultural development will likely emphasize the pattern already begun in the preceding decade," and that this strategy would be likely to benefit farmers with middle and higher incomes, leaving the great majority of small farmers to the attentions of the Rural Poverty Eradication Program of the Fifth Plan.

It is therefore pertinent to ask at this point why or whether, given the economic importance of the agricultural population and their palpably disadvantaged position within the society and economy, either governments have not been obliged to give greater political priority to improving their position or indeed why or whether rural people themselves have not set their own political priorities higher. To put it bluntly, governments seem to have been getting away, for a long time and to their advantage, with neglect or worse of the rural population; and for long the rural people seem to have been putting up with it to their disadvantage. Another way of asking these questions is to consider what are the various forms of labor control and, more generally, the forms of social discipline, representation, and participation, through what social means they have their effects, and what those effects are. The forms of economic differentiation, fragmentation, and convergence of relations of production, social relations, and interests considered in this volume are not wholly determined by economic or agronomic factors, nor do they have unambiguous social, political, or ideological consequences. They are conditioned and constrained by the latter as well as influencing them in turn. Their interpretation must include an analysis of wider contexts of the exercise of social power. This is attempted in the two papers which follow, though each has a different emphasis. Some broad features may be anticipated here.

First we have to note—in comparison with the other countries represented in this volume—the relatively high degree of religious, linguistic, and ethnic homogeneity of the country, the absence of popular anticolonial or successful revolutionary struggles, the historical continuity of ruling institutions, and the degree of geographical and political coordination and centralization of the country. We may note the absence of an established political party system with enduring local forms of organization and participation. We may also note the existence of laws which expressly forbid the formation of rural organizations of a trade union type, as well as laws and authoritative practices which suppress forms of political organization and dissent.

The rural majority does have an explicit political importance for governments—for example, in periodic though irregularly held elections and in the rhetoric of 'national stability' and 'anticommunism.' But votes can be bought, traditional loyalties (nationalistic, religious, local, personal) can be called on (although they can also have a reverse effect mitigating extreme manifestations of control), and the considerable coercive powers of state and allied agencies can be mobilized. They can and they are; and in order that they can be, the role of 'local powers' is crucially important: the local bureaucratic, commercial, and 'developmental' beneficiaries of the state and state policies.

For the benefits of the involvement of state and capital in the countryside have trickled down to, or have been tapped by, a sizable constituency which includes or coincides with those overlapping percentiles—5, 10, 15% or so—of producers effectively reached by the various support schemes, policies, and forms of organization dealt with earlier. This is not to suggest, however, that state patronage is the only or principal source of their *economic* strength. One might say that they represent the political constituency of governments in the countryside and that they, or some of them and in various alliances, extend the work of the state in policing, by proxy so to speak, the majority of others. For example, this nexus of local powers was probably a major external factor in the defeat of the one recent attempt by poor farmers at large-scale mobilization, the Peasants' Federation of Thailand (PFT) of 1974–1975 (see Turton 1982 and Anan, Chapter 5). Though short-lived, the PFT manifested something of the potential extent of 'peasant' opposition to political and economic conditions in the countryside—providing evidence, if such were required, which makes it impossible to continue to speak of the indifference or inability of small rural producers to associate and find a political voice. These local powers are integrally involved in increasing state 'penetration' of village political and administrative process (see Hirsch 1985), which is often officially subsumed under the category 'development' but which includes a proliferation of forms of social surveillance and control. And while 'democracy' and 'participation' are watchwords of policy, government strategy seems in effect somewhat to resemble the Indonesian New Order state's 'floating mass' strategy of separating the rural population from organized political activity.

In spite of such an array of central government policies and instruments for maintaining a political status quo (while in fact altering it), for ensuring uniformity of control and stability, in the past five years or so there have emerged a surprising number and variety of local initiatives and critical analyses of official development policies and programs. Some of these are more economic, others more ethical-religious, in purpose; some are rather nostalgically communitarian, others possibly looking towards more class-based forms of association. Though the constraints on such initiatives are enormous, some new ideas and groupings may emerge, coalesce, re-form, and contribute to the formulation of alternative rural strategies. The fifth five-year national economic and social development plan also aimed to encourage 'participation' and a greater role for local-level institutions, especially the subdistrict councils, to take locally relevant decisions on development issues. While this is likely to have strengthened the hand of existing local powers, it has yet to be seen what opportunities it may have presented for more effective participation by the majority of hitherto excluded sectors of the rural population.

Note

1. An earlier version of this chapter appeared in Turton (1987).

References

Douglass, M. (1984). *Regional Integration on the Capitalist Periphery: The Central Plains of Thailand.* Research Report Series, no. 15. The Hague: Institute of Social Studies.

Far Eastern Economic Review (1984). *Asia 1984 Yearbook.* Hongkong: Far Eastern Economic Review.

Hawes, G. (1982). "South East Asian Agribusiness: The New International Division of Labor." *Bulletin of Concerned Asian Scholars* 14:20–29.

Hirsch, P. (1985). "Village into State and State into Village: The Rural Development Entry." Paper presented to the Third Thai-European Seminar on State and Village in Thai Studies, Thai Khadi Research Institute, Hua Hin, Thailand, 9–12 April 1985.

Ingram, J. (1971). *Economic Change in Thailand 1850–1970*. Stanford: Stanford University Press.

Omvedt, G. (1986). *Women in Popular Movements in India and Thailand During the Decade of Women*. Geneva: UNRISD.

Panayotou, T. (1985). "National Policies Towards the Agrarian Sector in Thailand." Paper presented to the Social Science Research Council conference on National Policies Toward the Agricultural Sector in Southeast Asia, Chiang Mai, January 1985.

Pasuch Phongphaichit (1980). "The Open Economy and Its Friends." *Pacific Affairs* 53:440–460.

——— (1982). *From Peasant Girls to Bangkok Masseuses*. Geneva: International Labor Office.

Somsakdi Xuto et al. (1983). *Strategies and Measures for the Development of Thailand*. Bangkok: Thai Universities Research Association, NIDA, and Friedrich Ebert Stiftung.

Suehiro, A. (1985). *Capital Accumulation and Industrial Development in Thailand*. Bangkok: Chulalongkorn University Social Research Institute.

Thailand (1963). *1963 Agricultural Census Report*. Bangkok: Office of the Prime Minister, National Statistical Office.

——— (1973). *Land Use in Thailand*. Bangkok: Ministry of Agriculture and Cooperatives.

——— (1976). *Landholdings of Agricultural Households in Thailand*. Bangkok: Ministry of Agriculture and Cooperatives, Agricultural Land Reform Office.

——— (1978). *1978 Agricultural Census Report*. Bangkok: Office of the Prime Minister, National Statistical Office.

——— (1984). *Agricultural Statistics of Thailand Crop Year 1983/4*. Bangkok, Ministry of Agriculture and Cooperatives.

Turton, A. (1982). "Poverty, Reform and Class Struggle in Rural Thailand." In S. Jones, P. C. Joshi, and M. Murmis, eds., *Rural Poverty and Agrarian Reform*. New Delhi: ENDA and Allied Publishers.

——— (1987). *Production, Power and Participation: Experiences of Poor Farmers' Groups in Thailand*. Geneva: UNRISD.

Witayakorn Chiengkul (1982). "The Effects of Capitalist Penetration in the Transformation of the Agrarian Structure in the Central Region of Thailand (1960–1980)." Master of Development thesis, Institute of Social Studies, The Hague.

World Bank (1978). *Thailand: Selected Issues in Rural Development.* Background Working Paper No. 4 for *Thailand: Towards a Development Strategy of Full Participation.* Washington, D.C.: World Bank.

——— (1980). *Thailand: Coping with Structural Change in a Dynamic Economy.* Washington, D.C.: World Bank.

——— (1983). *Thailand: Rural Growth and Employment.* Washington, D.C.: World Bank.

Chapter Four

Local Powers and Rural Differentiation

ANDREW TURTON

This paper considers in an exploratory way some aspects of state and class formation. The main effort is to conceptualize the principal social forces engaged at the local level and their relations not only with overtly state structures and policies but also with what may be seen as less formal local manifestations and supports (and possibly distortions) of state powers. The main conceptual focus is on what I refer to as 'local powers,' located empirically within villages, within district and higher-level towns and centers, and within agricultural, bureaucratic, and commercial spheres and institutions. I shall emphasize the hybrid class and nonclass composition of these local powers, and examine their connections with the equally hybrid and heterogeneous power bloc of the state itself.

Contemporary Thai society, and more especially its rural sectors, cannot be adequately grasped simply in terms of a dominant mode of production. Perhaps no society can; in any case the results of such an attempt would be likely to be all the more inadequate in a situation such as we have here—where changes are recent and rapid and where large sections of society are not directly involved in dominant production relations, or only recently involved and in ways less clearly determinate than, say, a classic bourgeoisie or proletariat. There is therefore all the more reason to attend to the political-ideological fields of struggle in which class and nonclass subjects and identities *become* defined and developed, socially, politically, and ideologically. With such an approach we may, for example, be better able to assess the proposition sometimes made for Thailand and comparable countries (e.g., Indonesia): that a 'people/officialdom' or 'people/power bloc' contradiction is currently preponderant over a classic class contradiction. And if "class struggle is first of all a struggle about the constitution of class subjects before it is a struggle between class subjects . . . it follows from this that the field of political intervention is extremely broad" (Jessop 1982:196). And again, if as Jessop (1982:243) puts it, "the indeterminancy of class forces in relation to class location provides the space for the practices involved in securing hegemony" (or attempting to extend or challenge hegemony), then this clearly has broad implications for policymaking and political strategy, whether of currently dominant or subordinate groups.

Readers will be familiar with repeated calls for analyses of concrete historical situations which combine 'micro' and 'macro' perspectives and acquainted with the difficulty of the task and general paucity of response. The approach here tries to relate conceptually intensive village studies on the one hand and statistically based or more abstract society-wide studies on the other. To a lesser extent there is an attempt to bridge these two kinds of study empirically; though at the same time there is an effort to overcome distinctions of institutional or spatial 'levels.' Inevitably the attempt leads to a degree of synthesis and generalization and gives this paper a status somewhere in-between the general arguments of Part I and the more particular village studies which follow.

The paper draws on some recent empirical research, especially by Thai scholars which, while still limited in quantity and geographical scope, has begun to mark a new round of excellent scholarship and analysis. This includes intensive rural field studies and studies of urban and industrial sectors.[1] Data are especially drawn from a multidisciplinary research project which included a number of local but not exclusively single-village studies in rural areas of Northern (Chiang Mai, Uthai Thani provinces), Northeastern (Kalasin, Nakhon Ratchasima, Udorn, Yasothorn provinces), and East Central (Chachoengsao province), regions of Thailand.[2] The studies were carried out in an extensive, reflexive, and participatory style by a team which combined local activists and academic researchers, including the authors of both the papers on Thailand in this volume. The project took place in the period 1981–1982 but referred to conditions and experiences from the mid-1970s; and in all localities at least one member of the research team had lived for some years prior to the project. Most localities were in rice-growing regions or included rice-growing villages, but they varied in the extent to which other crops were grown. While forms and problems of production varied between case studies, a remarkable degree of similarity was found in the broadly political contexts of production. This in itself seems to warrant a degree of generalization in the presentation of the material. The conclusions and conceptualizations prompted by the ethnographic data will of course have to be tested in other localities (rice-growing or not) and in the light of changing material circumstances and theoretical perspectives.

Theoretically this paper draws on some currents of poststructuralist thinking about power, discourse, and subjectivity and on some 'neo-Gramscian' and 'neo-Marxist' thinking about state, class, and ideology (for example, Foucault 1980; Giddens 1981; Jessop 1982; Laclau 1977; Mouffe 1979; Therborn 1980, 1984; see also Turton 1984). The main aim is to contribute to further conceptualization of some of the issues raised in this volume in order to provide methodological suggestions for further empirical research and to stimulate debate and counterargument.

State and Agrarian Trends

The period with which this paper is concerned begins with the self-styled 'revolutionary' intervention of Field Marshal Sarit (1958–1963) and adoption of a World Bank development strategy in the first national development plan in 1961. In most respects there was no revolutionary break at all. Compared with other Southeast Asian countries, Thai society appears to be characterized by far greater historical continuities.[3] Sarit perpetuated and strengthened the controlling influence of the military, which has been the preponderant political factor in the state for the past fifty years. He inherited a (still) largely unreconstructed bureaucracy. For most of the period of Sarit and his successors (1958–1973) the constitution was abrogated, parliament dissolved, and rule by decree and martial law prevailed. Sarit strengthened his and the state's authority by reinforcing or reviving the role of the monarchy and tradition and by his style of 'despotic paternalism' (Thak 1979).

Sarit also, it seems, invented a new notion of development (*kan phattana*) the monopoly of which, politically and ideologically, lent legitimacy to an authoritarian state, provided material support for ruling apparatuses, created at least an impression of a concerned government (cf. Krinks, in Lea and Chaudhri 1983), and justified and was materially combined with an overriding concern with political stability (cf. Robison 1982:53).

Economically the Sarit years, following 'an era of haphazard, state-led development' (Hewison 1985:275), inaugurated a more profound break. Political and economic interventions by the state have played a substantial role in creating the conditions for capital accumulation by an indigenous bourgeoisie, though this has largely been outside agriculture (see Chapter 3). This is a fractionalized bourgeoisie, with a dominant banking and industrial fraction. There is a heavy concentration of ownership of capital by a small number of groups and families (Saneh 1983:29; Krirkiet 1982; Suehiro 1985). The largest group is that of the Bangkok Bank, controlled by one family, which has extensive interests in the rural economy not only through banking and finance but also through agroindustries, marketing, import-export, land, and services (Hewison 1981). This new bourgeoisie includes some members of the senior military and former military, high-level bureaucrats, and royal aristocratic strata. There is a complex overlapping and simultaneity of fundamental socioeconomic relations (as for rural producers) not only among many members of various bourgeois fractions but also with members of ruling and governing classes as such. When this combines with such social factors as common background, kinship and marriage, education, conferment of elite status through royal patronage, multiple connections of single families, and so on, this contributes to alliances which form part of the configuration of social forces we may term a ruling 'power bloc.' This degree of convergence of the political interests of these class strata (and with international classes) was evident in the events of 1975–1976 (Anan, Chapter 5).

However, within the power bloc the military continue to be the dominant element, a military whose role has historically been almost entirely one of internal consolidation (Anderson 1978:203). The bourgeoisie as a class does not effectively control the state and its apparatuses, despite the many close connections and influences. Nor does the state exist primarily to serve the interests of the bourgeoisie. While it does indeed serve their interests, it also has a momentum of its own and in large part serves its own somewhat hybrid, parasitic, 'secondary complex of predatory interests' (Thompson 1978:50), which constitutes much of the nonbourgeois elements of the power bloc and its local supports, appropriating wealth in often noncapitalistic and sometimes coercive and illegal ways. This suggests a specific kind of relative autonomy of the state or a connection between economic and political-ideological dominance. I hope to show that this is particularly relevant for the analysis of the local manifestations of state power or powers which combine with and also perhaps have a degree of autonomy from state power or represent competing factions within state apparatuses.

I am arguing that it is not possible to consider local powers without some attempt, however provisional, to characterize the nature of the state and that indeed such a characterization would be incomplete without the local focus. In order to complement this very brief presentation the reader is referred to the concluding sections of the preceding paper: 'Government Policies Towards Agriculture and the Rice Sector' and 'Political Aspects of Government Policies Towards Agriculture.' State supports seem to be having the effect of differentiating agriculture into two sectors: one advanced, in which agrobusiness and large-scale farming (especially of export crops other than rice) using modern technology is favored, and one backward and disadvantaged, which may at best receive the attention of 'poverty eradication' or 'job creation' programs (see below). One might say that in-between, and in some ways linking the two, are a minority of producers (though not exclusively producers, but those who are also officeholders, traders, etc.) who seem to be clearly net beneficiaries of limited state economic patronage and, directly or indirectly, state political patronage. And it is often by and through this social stratum that state and parastatal powers are exercised towards the majority of the poorer and disadvantaged rural population. The main part of this paper deals with these minority local powers, but first some categories of differentiation of the majority are examined.

Differentiation, Fragmentation, Convergence

I have argued elsewhere (Turton 1984) that not only are small owner-producers of various kinds increasingly drawn into commodity circuits in the sphere of exchange, and increasingly unable to reproduce themselves independently of these circuits, but that also and thereby they are decreasingly in control of strategic means of production and labor processes. That while *seeming* to own or possess their most basic means of production—land and labor (including its health, skill, etc.)—even this control is undermined and new factors over which they have no control become increasingly preponderant (including finance, inputs, technology, information, decision-making, transport, processing, storage, etc.). It might from this perspective be argued that a new class of rural producer has been brought into being: 'wage-labor equivalent' producers of surplus value (Banaji 1977; Bernstein 1982:176). And yet at the same time many new divisions have been created, by branch of production, forms of labor process, etc., which fragment and cross-cut lines of class, community, and household. Even the individual producer may be simultaneously landowner, sharecropper/tenant, wage worker, hirer of labor, petty trader, etc. In such a situation it may be premature to attempt to define exclusive categories of classes of rural producers. In any case such definitions become problematic when we consider questions of social consciousness. If we are to avoid mechanistic, static, teleological, or dogmatic forms of analysis, and in a concrete situation of relatively recent and rapid changes, then we are bound to say that 'class formation' and all the more 'class consciousness' are in an inchoate and fluid state. And nonclass or relatively non-class-specific factors (for example, new and old forms of national, regional, local, personal, and religious loyalties) are often predominant in the political and ideological arena in which in the course of struggles classes become defined and developed.

The calculus of the degree of objective and perceived common interests of different strata—in what issues, in which situations, and the extent of possible coalition, common discourse, and political practice—is dauntingly complex to imagine. Existing statistics, case studies, regional reports, etc., scarcely permit an adequate global economic analysis of trends, processes, and mechanisms of agrarian differentiation. Here I suggest an approach from another perspective: a critical review—a deconstruction if you like—of some current statistical, social, and ideological categories and criteria of differentiation and their political and ideological implications or effects, for example in the formation of social subjectivities.

At a most inclusive level there are in political discourse ideological categories of unity and solidarity—for example, of 'one nation' (e.g., *chao thai* subsuming or suppressing the distinctions of regional identities, non-Thai and non-Buddhist minorities, etc.) and 'one people,' 'the people' (e.g., *phrachachon*). Less inclusively there is the category 'villagers' (*chao ban*) connoting an undifferentiated unity of communities or the rural population in general (though I doubt whether some residents of rural villages would in all situations wish to identify themselves as such). Even more specifically there are the categories *chao rai chao na* ('farmer,' 'peasant'), which may occasionally be used of and by those who in fact possess no land, and *kasetakorn* (also 'farmer,' 'agriculturalist') used especially in official discourse, which may include various categories of large-scale commercial agricultural landowners and interests who are scarcely *chao rai chao na*. There are of course many other such categories and identities and I can do little more than suggest here that their use and effectivity in various forms of discourse deserve much greater attention (see also Turton 1984).

Seemingly more concrete categories such as landowner (especially 'small farmer'), tenant, wage laborer, and the rural poor need to be examined for their problematic assumptions and variety of uses. For example, over the past twenty years there has been an increase in the number of large and very large agricultural landholdings and an increase in the number of the landless agricultural and rural population. But proportionately more significant and more problematic for an analysis of agrarian trends is the persistence, or perhaps reconstitution, of large numbers of small to middling size landowners, a majority of the rural population. It is these I mainly consider here. Many policy conclusions and ideological interpretations are drawn from the existence of so many 'small farmers,' which both underestimate the degree of differentiation among them and at the same time the precariousness of livelihood and degree of commonality of situation, problems, and interests shared by different categories of 'small farmer.'

The precariousness of livelihood of the 'small farmer' and the potentially ideological nature of some uses of the term are indicated by the situation in the Northeast, which contains some 40% of all agricultural households (1978) and where the great majority of agricultural households have access to some land and 89% of farm operators are owners (World Bank 1978:34) and yet where over 70% of rural households could be classified as 'poor' or 'nearly poor' (under 200 baht monthly—approximately US$8—per capita income).

Available land statistics, which are for size of operated farmholding rather than ownership, are of limited value as indicators of differentiation. Consider these situations reported in a recent study (Turton 1987). In a village in Kalasin (northeastern region) the average (and near modal) holding was approximately one hectare, with the majority of households producing tobacco for a transnational company as their main or sole cash crop on at most one-third of their holding, the rest being planted to rice. In a bad year (1977), 80% of households produced insufficient rice for domestic consumption. In neighboring Yasothorn province, the membership (351 households) of a particularly autonomous farmers' association (who were opposed to the interests of the minority upper stratum of the village as discussed below) had an average (fairly evenly distributed) of approximately 5.5 ha (34 *raï*) of land per member, much of it planted to rice, i.e., in the top 25% of landholding size in the Northeast. But here low yields, family size (average seven members compared with average household size in the Northeast of under five for the statistical category 'well-to-do'), and pervasive drought, among other factors, meant that even this amount of land did not guarantee security of livelihood. In a Chiang Mai basin village (northern region), a holding of approximately 1.6 ha (10 *rai*), irrigated and double-cropped, indicated a 'rich farmer' in villagers' own terms, while the 'poor' (approximately 0.3 ha or under 2 *raï*) and 'middling' farmers (approximately 1 ha or 6 *rai*, again using villagers' categories), who together comprised 76% of households, were much more likely to possess unirrigated, more distant land, of lower quality and yield, and to have to rent in land, the poorest being the least able to rent from kin-related households.

The categories 'tenant' and 'part-owner/part-tenant' similarly conceal a variety of positions. Especially in rice-growing regions tenancy is widespread and growing (see Chapters 3 and 5). The figures vary partly over time and partly according to the areas selected. Thus in 1973–1974 in Chiang Mai province as a whole there was a reported total of 3.9% full tenants and 35.6% part-tenants; in the late 1970s over 50% of households in the (mostly) irrigated *valley* areas (i.e., rice-growing) were tenants or part-tenants, some 23% of them *full* tenants (Chayan 1984). Tenancies vary according to crop, kind and amount of rent (share, cash, preponderantly cash in the central region), and according to the nature of the relationship, which may be one of familial tenancy, tenants of poorer owners, 'noncapitalist' landlords, or of capitalist landowners (investing in and outside the agrarian sector). In the central region, in rice-growing areas, it is especially the larger capitalist farmers who are likely to be part-tenants and the 'marginal' producers full tenants (Douglass 1984:90).

Processes of differentiation and fragmentation may at the same time place a variety of categories of small landowning producers in a common situation, at least in some respects. These processes may be considered more dynamically by looking at factors which contribute to the dispossession and/or increasing marginalization of these producers.

Indebtedness and inability to repay at high interest rates (Douglass 1984:82, 84) appear to be the major precipitating factors in the dispossession of the small landowner or in stages of dispossession leading to foreclosure or sale of land. Of the contributing factors—which may be variously combined and some of which may in some cases become precipitating factors—the most frequently attested to by producers themselves are probably low market prices for agricultural products and high input prices. When such inputs are unavailable or unaffordable, this too reinforces the cycles of low and uncertain productivity, the squeeze effect of the 'simple reproduction squeeze' (Bernstein 1982). Prices to the producer are further depressed by preharvest contracted sale of crops or immediate postharvest sale; by various changing quota and grading systems; and by deductions for storage, transport and other costs, delayed payment, and so on. In some situations and with some crops the producers may not even be able to sell at all—all the more serious when the producer has switched to nonfood or nonstaple crops.

Health costs, in both cash and labor, are a major factor in the reproduction of household labor. Health crises, chronic and debilitating sickness, are frequently precipitating factors. They are directly related to the intensification of work, decline in levels of consumption, working in recently cleared forest-fringe areas with additional health hazards (e.g., fevers), injurious effects of toxic chemicals, etc. Other more routine expenditures—which may in combination take on a critical dimension—include retail costs of daily consumption (including purchased rice even for rice-producing households), educational costs, payment of fees, fines, and other compulsory contributions, legal and illegal.

Land may be lost directly through state organizations (e.g., the military) reclaiming land historically rented out or regarded as effectively possessed by the producers; or by forestry schemes, construction of dams, roads, and other 'public' works—with or without compensation. Land value and fertility may decline as a result of pollution and the effects of infrastructural developments. Variable legal security of tenure may compound the problem; the smaller landowners are less likely to have any legal tenure at all.[4] The consequences of partible inheritance for the fragmentation of smallholdings, hitherto not marked, may possibly be an increasingly relevant factor as the extent of undeveloped and easily developed land decreases, at least for some household members where fragmentation is resisted. The regressive land tax is a direct factor in the consolidation of larger landholdings, but it is inapplicable or low for small owners and less perceived as a problem.

In addition to factors leading to dispossession there are others which decrease or deny access to alternative opportunities. For example, the possibility of moving to other land in peripheral rural areas may be restricted by enclosurelike policies (e.g., preemptive purchase of large tracts of forest reserve land by investors with political influence). Opportunities for taking up tenancies are limited by increasingly prevalent forms of premium, and agricultural wage labor is reduced by mechanization. Access to 'free' productive resources (e.g., animal and vegetable resources of woodland and water) for subsistence or commodity production are diminished by deforestation.

As noted in Chapter 3, estimates for completely landless rural households (from 10 to 25% or so) do not reveal the number of members of small landowning households who are effectively landless and engaged in wage labor (locally, or as cyclical migrants, etc.). Nor do they give an adequate picture of the extent of wage labor or the extent to which hiring labor is a capitalist relation.[5] Nor do they indicate the extent to which wage labor is for agricultural purposes; though a recent World Bank report (1983) indicates that some 40% of the income of agricultural households is from nonagricultural sources. Consideration of figures for unemployment and underemployment give some further idea of the effects of the 'simple reproduction squeeze' on small producers. In 1979 two-thirds of members of landowning households in the Northeast, under one-third in the central region, were virtually unemployed for at least three months of the year after harvest. Of those employed many are reckoned to be 'underemployed': in the Northeast 42% in 1975 and 46% in 1976 (Kosit et al. 1981). Of those seasonally unemployed ('waiting for the agricultural season') two-thirds to three-quarters are women, and of the men most are young (50% under twenty-five, almost two-thirds under thirty) and single (over 75%). The World Bank, in reporting these labor reserves, is not optimistic that they are a resource 'that could be tapped at low cost' (World Bank 1983:92). Much political capital has been made of various 'job creation' or 'local development fund' schemes since 1975, and the fifth National Development Plan (1982–1986) claimed that 3 million of the 7.5 million 'rural poor' should benefit from employment creation schemes under the Poverty Eradication Program. These figures are just for those in selected 'target areas,' which include about half the rural districts in the North and Northeast, distributed in almost all provinces, and in some areas of the South, but none in the central region—a total of 246 districts in 37 provinces (NESDB 1981). Some idea of the tokenism of such schemes may be gathered from the figure of eleven days worked on average by 1.5 million people under work creation schemes in 1975 and 1976 (Kosit et al. 1981; see also Anan, Chapter 5).

The term 'rural poor' (or equivalent) is frequently used not only by outsiders but also by members of the various categories of agricultural producer discussed above. As a term of self-identification, in its political, social, cultural, and for that matter also economic content, it refers to more than the official income categories 'poor' and 'nearly poor.' It includes such categories as 'landless,' 'poor peasant,' 'marginal peasant,' and many who might be categorized as 'middle peasant.' It refers in some directions to more, and in others less, than the category 'small farmer' in vogue in official and international development circles. While imprecise and of limited analytical value, it at least avoids—not least in the perceptions and tactical consideration of some peasant leaders—some of the pitfalls of conceptual classification that is inappropriately precise or premature (whether due simply to lack of data, inadequate concepts, or possibly to lack of 'resolution' of complex and recent changes).

Subordination and exploitation are frequently expressed by the rural poor as 'disadvantage,' where those who 'gain advantage' (*dai phriap*) at the expense of the poor are identified as those with social power, privilege, and influence, as well as controlling economic power through specific forms of appropriation and control of production and means of production. 'Disadvantage' also connotes a wide range of social and human disparagement and loss of dignity. Analysis of relations of production has to take into account these other forms of social differentiation and consciousness of them. Many poor rural producers are keenly aware of the economic dominance of social forces located at some geographical and social distance (banks, agribusiness companies, and other national and transnational agencies and classes) and also of the political dominance of higher-level state, military, and bureaucratic agencies. But their immediate relations are with various local economic and bureaucratic agencies and local power blocs which mediate the intervention of dominant forces in society. These local powers are themselves by no means undifferentiated, but there is a sense in which their global relationship with the rural poor constitutes a major contradiction (analytically as well as phenomenologically) and site of struggle. It is to these local powers that I now turn.

Local Powers

An empirical distinction might be made—though as we shall see it is increasingly problematic—between a 'village level' rural sector, up to the subdistrict level, and a 'higher' level including the district centers and linkages to province and regional levels and beyond. To an extent this is a distinction which corresponds with a growing contradiction between town and country in agricultural areas, though empirically some subdistrict centers may be small towns and some district centers little more than villages with a bureaucratic superstructure. There is increasing 'traffic' between these two sectors, but compared with some countries (e.g., Sri Lanka) locally dominant strata are less likely to be actually resident in villages. I wish to characterize those who exercise political and economic power at each level, their links with categories of rural producer, and their 'vertical integration' with state and capitalist structures and milieus.

At the 'village level' we find, in village after village, a small minority—perhaps some 5% more or less—of households which possess a degree of wealth, control of resources, prestige, and power which sets them apart from the majority. These correspond statistically with the 'well-to-do' income category (e.g., about 7% for the Northeast), though there are concentrations in some villages and many 'satellite' villages consist entirely of poorer producers. While many of these households or families were already well placed at the beginning of the period under consideration, there is evidence that many formerly well-established families have not maintained their dominant position—by not taking opportunities to invest in production or in the education of children, through the fragmenting effect of inheritance on large holdings, and so on.[6] The forms of 'primitive accumulation' before and during the present period are many and varied; and they undoubtedly include illegal means. Many are specific to the recent past and involve varying degrees of direct and indirect state patronage. Among these are privileged access to low-interest credit and limited inputs provided on credit; access to state-owned land (e.g., establishing plantations in forest reserve areas); access to contracts to supply quotas; access to limited official budgets for development projects (e.g., free inputs for trials, for 'model' farmers, etc.) and guaranteed prices (e.g., through privileged position in official farmers' associations); and access to the benefits of infrastructural development.

It seems likely that a substantial proportion of the currently dominant village strata is now increasingly able to maintain and reproduce itself as part of a small capitalist class, though such a position does not fully or unambiguously define their social location and identity. At the village level they include—and often in the same person, certainly family—larger landowners, commodity dealers, shopkeepers, village officials, some teachers, rice millers, moneylenders, owners of small-scale transport and machinery, large-scale employers of wage labor, etc. They derive advantage from their external connections and alliances and from their roles in 'linking' the majority of villagers with state and market structures, above all through relations which enable them to accumulate (or be the first stage in the accumulation of) village surplus through wages, commodity dealing, retail prices, rent, and interest. In this they are crucially dependent both on the external capitalist sector and on the state-bureaucratic sector. In relation to the latter they are the beneficiaries of various forms of state patronage. Economically this includes not only access to credit, agricultural inputs, and quotas, as mentioned, but also salaries, fees, stipends, per diem allowances for attending meetings, proportions of official budgets, political handouts, strategic economic information, and preferential access to education and health services. In a more political sense they receive offices (which in turn allow for all sorts of nonofficial remunerative opportunities); access to higher-level powerholders and the coercive apparatuses of the state, the right to possess weapons; state-backed authority, legitimacy, and prestige; and support for judicial and extrajudicial decisions and actions.

Within the village they are thereby enabled effectively to control and monopolize state-initiated committees, associations, development projects, and so on. Concurrently they can reinforce their politically and economically dominant positions through various forms of noneconomic patronage (social, religious, etc.) which allows them to benefit from association with more traditional and village-derived forms of relationship and legitimacy, and in ways which are often no longer reciprocal, or not to the same extent as formerly, and yet can have the ideological effect of mitigating or concealing the extent of differentiation and contradiction.

The wealth and power accumulated in these ways (both capitalist and 'noncapitalist') are converted into further forms of wealth and power. Probably what is most prevalent is a combination of unambiguously capitalist investment and accumulation (both inside and outside the agricultural sector) with expenditure, investment, in ways which are less directly related to their reproduction as a distinctly capitalist kind of class stratum. For example, if we consider patterns of consumption, we can certainly find instances of individuals who eschew lavish expenditure and concentrate on classic small capitalist accumulation and investment. More commonly we find large sums being spent (mainly outside the village economy) on houses, vehicles, and (largely imported) consumer 'durables' of every kind (which are available on a prodigious scale throughout the country), on travel, and on social events such as weddings, domestic ceremonies, feasts, drinking parties, and gambling. Much of this is geared to establishing and reproducing social relations with strategic superiors and subordinates in order to enhance political and economic position, to secure lucrative offices and contracts, to gain protection for illegal economic activities, and to accumulate political and economic clients. This may be supplemented by straightforward payment of bribes (though the kinds and procedures may be anything but straightforward). Enormous sums are also spent on more direct campaigning for various elected or appointed offices. Other major items of expenditure are the procuring of educational and job opportunities for children (which may involve large cash inducements) and in maintaining children in secondary or tertiary education away from home.

Members of this minority upper stratum in villages are likely to have a multiplicity of economic, social, and political relationships with individuals and households, and with a range of them, from among the majority of the poor to middling villagers, whether or not these are *primarily* laborers, tenants, or small landowners or whether the relation is primarily in terms of one such position. As we have seen, economically this may be as employers, landlords, retailers, wholesale purchasers, moneylenders, 'quota foremen,' providers of services (e.g., milling, machinery hire), and as dominant members of farmers' associations, cooperatives, irrigation associations, credit associations, etc. Political relations include the formal administrative relation of village officials and those consequent on membership of committees, especially village committees and the subdistrict council (*sapha tambon*, see below), but also such groupings as Village Scouts and paramilitary 'volunteer defense forces' (see also below), all of which tend to be dominated at the village level by the stratum under discussion. Other available forms of patronage—for instance, in familial, religious, and educational spheres—have already been mentioned.

The forms of differentiation thus established by the reproduction of this stratum have the further effect of fragmenting and demobilizing the subordinate majority. This can occur through the discrimination and cooptation of selected 'clienteles' and the power to mobilize and divert activities (labor, time, money) in official directions or according to private interests. A further crucial effect of the above, backed by additional means of surveillance and intimidation (see Turton 1984), is the prevention and harassment of attempts to form autonomous organizations, whether by the poor or other specific strata and interests, which would tend to exclude the minority upper stratum (and of course their higher-level allies) from participation, let alone leadership and control (see Turton 1987).

Hall (1980:464), whose sources mainly refer to the situation up to the early 1970s, concludes his careful survey and analysis of 'the emergence of an elite peasant stratum monopolizing control of economic and political resources' by stressing the complexity of 'variation of intertwining and conflicting interests' between such an elite and 'rural bureaucratic and commercial classes' and the perceived 'precariousness' of relations between them. He finally suggests that 'to a very large extent, the great divide still lies between the peasant elite and the bureaucratic-urban world,' despite increased differentiation 'in opposition to other strata of the peasantry.' More than ten years on I would argue that both the 'internal' differentiation within villages and also the consolidation of the upper stratum is more marked; that it is no longer so appropriate to speak of a '*peasant* elite'; and that what I have so far provisionally termed a 'village-level upper stratum' cannot be conceptualized as a stratum distinct from much of the (local) 'supra-village' sector, to which I turn shortly. To anticipate, it may be that the 'great divide' is rather between the majority of poorer rural producers and a composite local bureaucratic/commercial class or coalition of class elements, which may be 'represented' by a local power bloc and includes and is supported by the village-level upper stratum.

I would like to enter some qualifications, however. First there are local officeholders (including subdistrict heads, *kamnan*) who in economic terms are members of poor or middle strata and who have identified their interests in common with them. A number of headmen were actively engaged in the farmers' movement of 1974–1975 and some 'went to the jungle' following those events (see Turton 1982). Secondly, there are instances of well-to-do farmers (some of them officeholders) who have acted in concert with poorer farmers—for example, on issues of agricultural prices and also in certain land dispute cases where they were perhaps fellow tenants or part-tenants (if larger ones) on land owned by absentee state or private landlords. In most cases of these two kinds known to me there were important noneconomic factors involved. For instance, there were regional cultural and geographical factors, sometimes linked with historical forms of greater village autonomy. Other cultural or ideological factors included various elements of transformed 'traditional,' Buddhist critiques of abuses of official and state power. Some of these have elements of a kind of 'Buddhist socialism,' while others seem to constitute an anticipation or yearning for a more modern, 'social-democratic' form and style of government, which would of course not be incompatible with the aspirations of an emergent rural capitalist class. In any case these suggest only temporary coalitions of interest.

Furthermore, a number of positions within the coalitions or alliances between village upper stratum and external bureaucratic and capitalist classes may be conceptualized. There are those who may have reached a point when they are no longer crucially dependent on state economic supports, in the immediate sense, but who may continue to rely for instance on the coercive apparatus of the state. There are those still crucially dependent on economic supports who, if these fail them, might be likely to sympathize with the thinking and actions of economically and socially lower strata. There are those too who without being economically so substantial may cooperate fully with local bureaucratic and capitalist classes. We can conceptualize various degrees of ideological and social commitment to such cooperation and alliance, varying with contexts and issues, and with the level of struggle and strengths of various contenders in local struggles to establish or challenge hegemony.

At the 'district level' are to be found, first, the representatives of the functional departments of various ministries, which have tended to assume and monopolize new tasks and extended their functional and administrative roles at local levels. These include the more autonomous police (characterized by Girling 1981:132 as a 'prebendal' police) and most importantly the district administration as such, headed by the District Officer and various deputies, among whom especially important in most rural areas is the deputy for 'defense and suppression' of the Internal Security Operations Command, in charge of paramilitary self-defense and other security groups and networks. There has been an increase in the involvement of the military at this level (Chai-anan 1982). Formally outside the bureaucratic sphere at this level are likely to be found large-scale counterparts of the village upper stratum: larger-scale landowners, merchants, millers, contractors, and owners of machinery and transportation. They are linked vertically 'upwards' with provincial and national levels (wholesalers, exporters and processors, national and transnational companies) and 'downwards' to their agents, suppliers, clients, and employees. Many will be related by kinship and other ties to the more junior and more permanently resident district officials. Their informal networks of friendship and association with district, subdistrict, and village officials and other notables constitute strongly influential groupings and mutual supports at this level. This does not of course rule out sometimes intense and violent interpersonal or factional rivalry and conflict.

Each district will have one or more elected representatives on the provincial assembly (*sapha cangwat*), which controls some budgets destined for use in the locality. One or more elected members of the National Assembly may be associated with a district, though this is not necessarily a parliamentary constituency. Not infrequently these representatives are drawn from commercial and bureaucratic interests referred to above. Landowning interests as such seem to be less represented. Their local economic agents are often their 'vote bosses' (*hua kanaen*) at election time. Personal funds of 1 or 2 million baht may be spent on campaigning. There is usually no formal local political party organization at these levels, which suggests a fluidity of political alliances and also the political importance of the networks and milieus described—at once politicizing commercial, administrative, and other less formal relations and contributing to the depoliticizing of the majority or at least excluding them in the short term from effective political processes.

I have elsewhere (Turton 1984:32) offered a composite profile of a member of the milieu under discussion, drawn, but I think not overdrawn, from a number of actual instances, this profile, if methodologically somewhat unorthodox, suits the unorthodoxy of the realities:

He is a provincial assembly elected representative, owns a transport company, and is involved in sales of agricultural inputs, owns some upland land currently being developed, and is engaged in construction work of all kinds (feeder roads, school buildings, police stations, rest houses, public meeting halls, small bridges, small dams and wells, etc.). He has various district and subdistrict officials 'in his pocket.' He receives some provincial funds for his 'territory,' from which he may take a percentage before passing it on and subsequently also profiting from contracts. He is a member of the local branch of Rotary (or Lions, Jaycee, etc.). He is also chairman of the local Village Scouts and a generous provider of funds for weapons for the local volunteer forces. He, and perhaps his wife, may also donate to local charities and royally sponsored projects (hospitals for example) for which they may hope to receive state honorifics. He, or more probably his subordinates, may likely call on 'gun hands' to do their dirty work for them. So he is involved in both capitalist activity and primitive accumulation, with much use of extra-economic coercion. If he were to operate on a big enough scale he might even be termed a local 'godfather,' a term which like 'mafia' has entered the Thai lexicon. Indeed the Ministry of the Interior itself, which controls a majority of local officials, is sometimes colloquially referred to as the 'ministry of the mafia.'

I do not wish to suggest that such figures, coalitions, and nexuses as I have described—symptomatic though they are—are uniform, universal, or monolithic, nor that they constitute impenetrable or unshakable bastions of power. There is also possibly an expansion of more technical-rational, universalistic, and legally organized spheres of both bureaucratic and capitalist activity: technocratic senior officials, recently graduated from the universities and posted to particular rural areas for short periods, larger wholesalers and dealers, large rice millers, bank officials (among the best-paid salaried workers), etc., who do not or have no need to operate in such ways or do so to a markedly lesser extent. Yet the frontline field agents of even the largest transnational companies are frequently to be found in and as part of the political milieu described. For example, 'quota heads,' foremen, subcontractors, privileged contracted suppliers, etc., are likely to be selected from among the village upper stratum. Company business may be transacted in the context of meetings not only of official farmers' associations but also those of paramilitary organizations—just one context of potential 'extraeconomic' coercion.

More generally, there exists a complex overlapping of economic, political, administrative, and cultural agencies, relations, and interests, along with a characteristic combination of formal and informal, official and nonofficial, public and private, legal and illegal activities and their 'horizontal' and 'vertical' linkages and forms of consolidation. Such local realities of power deserve greater theoretical prominence and conceptualization. They tend to be largely 'invisible' in much academic writing, perhaps because they defy formal analysis in existing paradigms, not to mention being a highly uncomfortable field research milieu. 'On the ground' they are not so invisible, though they are obscured by specific practices of secrecy and social exclusion. They are also evidenced from time to time in provincial and national newspapers, revelations which have cost the lives of numerous journalists. These realities, in their political and ideological as well as economic dimensions, are close to the conditions and experience of the rural poor; though there is probably more acute awareness of them among those who have attempted to form their own groupings and coalitions and have discovered for themselves the ramifications of local powers.

Through their hybrid class (and nonclass) composition these local powers are articulated with the hybrid and heterogeneous power bloc of the state itself. Neither wholly state nor private, and certainly not an 'apparatus,' they are one of the crucial means, direct and indirect, more or less consciously, and through a range of normative and remunerative sanctions backed by coercion, by which the state is able to prevent, divert, mitigate, and suppress the expression of conflict between rural classes and in a sense the very formation of those classes. In such ways, though by no means these alone, are political and economic interests of the state furthered and at the same time conditions created for further penetration of capital into the countryside.

Village-Level Organizations and Associations

Girling (1981:174) has reiterated a common judgment that 'the aloofness of Thai villagers from collective activities is in marked contrast to the more structured behavior of villagers in Vietnam, Sri Lanka, India, and elsewhere'; and he has argued that 'the virtual absence of permanent associations in most Thai villages . . . makes it difficult for the poorer peasants to organize effective opposition to the power of the rural elite, backed by the province officials.' Though the broad contrast with other countries has some validity, both 'aloofness' from collective activities and 'virtual absence' of associations may be questioned—and certainly the hypothesized relation between the two. To do so it is necessary to reexamine (if only briefly in the context of this paper) forms of village-level organization and association, some of which have already been alluded to. In particular we need to consider ways in which and the extent to which such organizations—or the lack of them—indicate, express, and actually serve to advance the interests and contribute to the formation of particular categories and strata of the rural population. We need to take into account ways in which old forms of association may have been undermined or transformed and new forms developed and *prevented.* We should also bear in mind the potentialities and effects of less 'structured' or 'permanent' forms of association.

There is not the space here to establish in any detail what these 'old forms of association' might have been. It would certainly be a mistake to exaggerate local or village autonomy and cohesion in earlier times or to establish some kind of immemorial baseline. We have to bear in mind the 'external' political and economic framework imposed, so to say, on villagers by the prevailing means of controlling village labor and produce in the nineteenth century and earlier (the '*sakdina* system') and the frequency with which village communities were disrupted, moved and resettled at the behest of officials and nobility. However, it can be shown that at least in the nineteenth century and often well into the twentieth there existed a degree of local independence, self-reliance (to use a term dear to certain development planners), cooperation, and relative egalitarianism within local communities, which present marked and increasingly stark contrasts with current realities. This was probably truer for the distinct regional cultural areas, for instance the North and Northeast (Chatthip 1984), and in more nucleated and long-established villages further from political centers and where there were no resident aristocrats or nobility (*caonai*). A number of studies provide evidence for what we might call forms of association of entire communities 'for themselves,' both for internal and external (intervillage) purposes of social reproduction, as well as more restricted interhousehold networks based on kinship and neighborhood (see de Young 1955; Kaufman 1960; Turton 1976; Sharp and Hanks 1978). These involved various forms of reciprocal and redistributive exchanges (for example, exchange of labor for agricultural and domestic reproduction tasks, pooling of labor for communal religious, social, infrastructural, and, in the North, irrigation tasks) and forms of collective decision-making and action (for example, related to the village temple, cults, and feasts; self-defense; avoidance, hue and cry, and hot pursuit of thieves, bandits, and predatory officials). In the latter, externally appointed village officials were likely to be predominantly village-oriented. Social differentiation based on age, gender, ritual rank, and *nakleng* ('strongman') status were of greater weight than, and differences of wealth less dependent on, differentiated relations of production or access to means of production.

When 'community development' policies were initiated in the early 1960s, there was official exhortation of 'community' solidarity and encouragement of 'community leaders' (often the trio: headman, teacher, 'abbot'). Yet ironically this coincided with, and underestimated and/or ideologically countered, a real undermining or destruction of such 'community' features by the very process of 'development': increased commoditization and socioeconomic differentiation within and between villages—and between villages and towns—and increasing bureaucratization of village, subdistrict, and district levels. Increasingly 'community leaders,' as viewed from the outside, are members of the upper stratum characterized earlier and are likely to be regarded by poor villagers as *klum ittiphon* (influential groups) whose 'society' (*sangkhom*), drinking circles, marriage patterns, and other restrictive means of socialization and (less visible, even secret) decision-making increasingly exclude the poor and *include* members of similar strata in other villages and from bureaucratic and commercial strata above the village.

In many respects former means of social reproduction are absolutely diminished—an example is the often heard complaint of not having enough time to hold community festivals. Probably more common is the apparent 'persistence'—alongside the newer, more exclusive forms of association for the upper stratum—of earlier forms but with a transformed content and significance, for example in religious, administrative, and even kinship spheres. Some older forms may be quite consciously transformed in a bureaucratized and commoditized direction—as in the increasing prevalence of contracted management of local irrigation systems in the North and even of religious festivals. I have already suggested that in such transformations the differentiation of the upper stratum can be simultaneously enhanced and concealed or mitigated in the perception of villagers. On the other hand, of course, it may be precisely in some of these spheres that a poor peasant leader may gain status of strategic value. Also to the extent that in given circumstances there is a temporary basis for a coalition of interests, then such relatively undifferentiated relations and forms of association may be of advantage to the majority.

Moreover there is evidence of the selective recovery and transformation of earlier values and relations in the discursive and organizational practice of the poor, especially those engaged in more autonomous efforts to work out alternative 'development' strategies in conjunction with external allies. While some such practices seem to be more consciously initiated by outsiders (an example perhaps is the more secular and redistributive use of almsgiving in Buddhist ceremonies), many more are likely to be initiated or consciously perpetuated by villagers themselves. These may include styles and occasions of meeting, discussion, visiting, and linkage beyond the village and also reappropriation of 'suppressed' traditions of critical discourse on domination and exploitation. The presence in the countryside of various categories of nonofficial development workers, agents of nongovernmental organizations, religious groups, and concerned professionals, mainly from educated, urban, and middle-class backgrounds, constitutes a new source of potential allies and interlocutors for the rural poor. Even if their numbers are few, their presence is quite widespread and possibly disproportionately effective.

Of all the more recently instituted forms of organization at the village level, the most pervasive and 'structured,' if not necessarily more 'permanent,' are those which have been officially introduced and are centrally controlled. They can be considered as being of three broad types: administrative, economic/developmental, and ideological/paramilitary. There is considerable overlap in personnel, policy, and function between these organizations. As argued earlier, they tend to favor the interests of the minority village upper stratum, though degrees of control and exclusion vary.

Probably the most important of the administrative structures is the subdistrict council (*sapha tambon*). This body began to take on a new significance and potentiality in 1975 when for the first time central funds were allocated to be spent on locally decided development projects. The 1982–1986 National Economic and Social Development Plan gave considerable emphasis and new powers to the council as part of a policy of 'decentralization' and 'participation.' In particular the councils were empowered to develop long-term economic and social plans for their localities. The council consists of the subdistrict headman (*kamnan*), a secretary (a teacher), village headmen and one other especially qualified or experienced person from each village, and a number of district-level officials *ex officio*. In practice the power of the subdistrict head is often such that he can control the nomination of members to the council and can determine the use of development funds according to privileged sectoral interests or official purposes (e.g., for paramilitary groups) other than those proposed by villagers or even decided in council. The greater powers of the council, while raising the stakes, may however offer some opportunities for new forms of political action by categories of villager other than the established upper stratum. It is pertinent to reiterate here that there are virtually no forms of permanent or even minimal political party organization at the district level and below.

The two main types of 'economic/developmental' associations are agricultural cooperatives (*sahakorn kan kaset*) and Farmers' Groups (*klum kasetakorn*). The former have been in existence for some sixty years and are administered on a district-wide basis by district-level officials. In 1979 only some 15% of agricultural households were members of cooperatives. The great majority are concerned almost exclusively with the provision of credit, and an official survey found the majority 'third rate' at best in terms of efficiency and variety of services.[7] Farmers' Groups were instituted in the 1970s after various earlier forms had been abandoned. These groups are quite similar to cooperatives in actual function and scale of membership, but they operate more at the village level with officeholders drawn largely from the upper stratum. Neither are encouraged or permitted to become too powerful, nor to extend horizontal linkages, nor to increase the bargaining power of members. While *de facto* excluding the majority of rural producers, they also *de jure* exclude the landless and many marginal farmers. However, in some instances various combinations of small farmers have been able to establish their own more autonomous organizations, using the legal format, and to a limited degree extending the potentialities of these official associations, adopting goals, styles, and procedures which are in marked contrast with, and often directly critical of, the practices and interests of the leadership of many such groups (Turton 1986, 1987).

The 'ideological/paramilitary' type of grouping includes notably various local militia or 'volunteer defense forces' and the Village Scouts, which combine, in varying proportions, ideological, developmental, intelligence gathering, surveillance, and security roles. These organizations might be regarded as being not very effective or continuous in practice, though they are at least in place for potentially greater mobilization. They do, however, contribute to the powers of coercion of locally dominant strata, to ideological notions of consensus and solidarity, and to definitions (backed by the Anti-Communist Activities Act) of popular initiatives as 'subversive,' 'communist,' etc. (see Turton 1984). They tend to have a fragmenting and demobilizing effect on potential or actual alternative forms of organization by subordinate strata.

There is not the space here to develop a critique of attempted explanations, whether in terms of 'aloofness,' individualism, tradition, or political apathy, of the relative lack of forms of collective organization by poor peasants. I have discussed elsewhere (Turton 1986, 1987) some of the peasant counterpoints to 'local powers,' forms of peasant resistance, resourcefulness, and creativity which lie in between 'everyday resistance' (cf. Scott and Kerkvliet 1986) and larger-scale historical mobilizations (cf. Kanoksak 1983; Chatthip 1984; Tanabe 1984). What has been argued here is that there is a complex local array of economic, political, and ideological structures, formal and informal processes, laws, regulations, and mentalities which constrain and limit (or may prevent or co-opt) the development of alternative or emergent ideas and practices on the part of specific economic strata or classes of the majority of the rural population.[8] At the same time they benefit and consolidate locally dominant agricultural, commercial, and bureaucratic strata whose articulation with state and ruling power bloc was considered earlier. There are also national-level factors which intervene to support the position of these local powers—such as, among others, the preponderant policy of national security and stability; restrictive legislation (including the laws on association which prevent *rural* workers from forming trade-union-type associations); and the weak development of parliamentary politics and political parties. A further question would be whether such factors also intervene to prevent members of this hybrid milieu from developing *their* more distinctly articulated and represented political interests.

Recapitulation

The conceptual and empirical focus of this paper has been on 'local powers'—power blocs and coalitions which are seen as a crucial mediation, nexus, and localization of contradictions and conflicts between state and capitalist spheres on the one hand and the majority of rural producers on the other. The heterogeneous or hybrid compositions of class and nonclass subjects at each of these levels has been given some emphasis, as has the 'indeterminacy of class forces in relation to class location' (Jessop 1982:243). There was an attempt to conceptualize sites and spaces of local struggle to secure or challenge hegemony and to constitute or prevent the emergence of particular forms of class and nonclass subjects and forms of consciousness and organizational practices. Within these spaces and struggles there was an attempt to conceptualize, and specify concretely, interacting determinations of economic, political, coercive, cultural, and ideological factors. These are of course effective within a specific historical social formation—in which we can discern a degree of relative autonomy of state and other political-ideological factors—but within limits set and altered by a particular kind of capitalist development, which in turn is partly and crucially determined by exogenous forces.

Notes

1. Rural studies include, for example, Anan (1984), Ananya (1985), Chayan (1984), Hirsch (1987), and studies of urban and industrial sectors (for example, Krirkiet 1982 and Suehiro 1985). Reynolds and Hong (1983) review much of the Thai-language debate on mode of production and social formation.

2. The Popular Participation in Rural Thailand project was conducted under the auspices of the United Nations Research Institute for Social Development. The research team included leaders and members of village-level associations of poor farmers and laborers, nonofficial development workers, and academic social scientists from Chiang Mai, Chulalongkorn, and Thammasat Universities, as well as the present author. I am heavily indebted to the collective work of all participants. Various publications based on the research have appeared and are summarized and developed in an overview report (Turton 1987).

3. See the preceding paper. Elsewhere, however, I have tried to show the ideological nature of many notions of 'continuity' and the degree of real discontinuity, conflict, contradiction, and restructuring of 'tradition' (Turton 1984).

4. In 1977 the proportions of agricultural land held under varying degrees of legal security of tenure were: Northeast 49%, North 63%, Center 83%, South 54%. Apart from the Center most of this land was not held with full legal tenure; in the Northeast the proportion was less than 5% (World Bank 1978).

5. In one central Thai village as early as 1970, while all 'commercial farmers' (7.2 ha) used hired labor, so did 83% of 'middle peasants' (2.6–7.1 ha), and 20% of 'marginal peasants' (0.16–2.5 ha) (Douglass 1984).

6. See, for example, a recent study in Chachoengsao province (Ananya 1985).

7. A Bank of Thailand report on agricultural credit in 1980 judged only

1.4% of cooperatives (14 out of 713 surveyed) to be 'first class' and 32.5% 'second class' in terms of efficiency and variety of operations (Witayakorn 1982).

8. This argument is developed and further substantiated in Turton (1987).

References

Anan Ganjanapan (1984). "The Partial Commercialization of Rice Production in Northern Thailand 1900–1981." Ph.D. thesis, Cornell University.

Ananya Bhuchongkul (1985). "From *Chao Na* (Farmer) to *Khon Ngan* (Worker): The Growing Divide in a Central Thai Village. Ph.D. thesis, School of Oriental and African Studies, University of London.

Anderson, B. (1978). "Studies of the Thai State: The State of Thai studies." In E. Ayal, ed., *The Study of Thailand*. Southeast Asia Series, no. 54. Athens, Ohio: Ohio University Center for International Studies.

Banaji, J. (1977). "Mode of Production in a Materialist Conception of History." *Capital and Class*. Vol. 3.

Bernstein, H. (1982). "Notes on Capital and Peasantry." In J. Harris, ed., *Rural Development: Theories of Peasant Economy and Agrarian Change*. London: Hutchinson University Library.

Chai-anan Samudavanija (1982). *The Thai Young Turks*. Singapore: Institute of Southeast Asian Studies.

Chatthip Nartsupha (1983). *Sethakit muubaan thai nai adit* [Thai village economy in the past]. Bangkok: Sangsan Publishing House.

——— (1984). "The Ideology of 'Holy Men' Revolts in North East Thailand." In A. Turton and S. Tanabe, eds., *History and Peasant Consciousness in South East Asia*. Senri Ethonological Studies 13:111–134. Osaka: National Museum of Ethnology.

Chayan Vaddhanaphuti (1984). "Cultural and Ideological Reproduction in Rural Northern Thai Society." Ph.D. thesis, Stanford University.

de Young, J. (1955). *Village Life in Modern Thailand*. Berkeley: University of California Press.

Douglass, M. (1984). *Regional Integration on the Capitalist Periphery: The Central Plains of Thailand*. Research Report Series, no. 15. The Hague: Institute of Social Studies.

Foucault, M. (1980). *Power/Knowledge*. Edited by C. Gordon. Brighton: Harvester Press.

Giddens, A. (1981). *A Contemporary Critique of Historical Materialism*. London: Macmillan.

Girling, J. (1981). *Thailand: Society and Politics*. Ithaca and London: Cornell University Press.

Hall, R. (1980). "Middlemen in the Politics of Rural Thailand: A Study of Articulation and Cleavage." *Modern Asian Studies* 14, no. 3:441–464.

Hewison, K. (1981). "The Financial Bourgeoisie in Thailand." *Journal of Contemporary Asia* 11:395–412.

——— (1985). "The State and Capitalist Development in Thailand." In R. Higgott and R. Robison, eds., *Southeast Asia: Essays in the Political Economy of Structural Change*. London: Routledge & Kegan Paul.

Hirsch, P. (1987). "Participation, Rural Development, and Changing Production Relations in Recently Settled Forest Areas of Thailand." Ph.D. thesis, School of Oriental and African Studies, University of London.

Jessop, B. (1982). *The Capitalist State*. Oxford: Martin Robertson.

Kanoksak Kaewthep (1983). "Bot wikhro sahapan chao naa chao rai haeng phrathet thai—kan khleuanwai khong chao naa yuk phrachathipatai berk baan" [Analysis of the Peasants Federation of Thailand: the farmers' movement in the 'Democratic Period']. *Sethasatkanmuang* [Journal of political economy] 2:1–44.

Kaufman, H. (1960). *Bangkhuad: A Community Study in Thailand*. Monographs of the Association of Asian Studies, no. 10. Locust Valley, N.Y.: J. J. Augustin.

Kosit Panpiamrat et al. (1981). *Chonabot Isan* [The rural northeast]. Bangkok: Thai Watana Phanit.

Krirkiet Pipatseritham (1982). *Wikhro Laksana Kanpencaokhong Thurakit Khanatyai Nai Phrathet Thai* [Analysis of the characteristics of big business ownership in Thailand]. Bangkok: Thammasat University, Faculty of Economics.

Laclau, E. (1977). *Politics and Ideology in Marxist Theory*. London: New Left Books.

Lea, D., and D. Chaudhri, eds. (1983). *Rural Development and the State*. London and New York: Methuen.

Mouffe, C. (1979). *Gramsci and Marxist Theory*. London: Routledge & Kegan Paul.

NESDB (1981). *Rural Poverty Eradication Program*. Bangkok: National Economic and Social Development Board.

Reynolds, C., and L. Hong (1983). "Marxism in Thai Historical Studies." *Journal of Asian Studies* 43:77–104.

Robison, R. (1982). "The Transformation of the State in Indonesia." *Bulletin of Concerned Asian Scholars* 14:48–60.

Saneh Chamarik (1983). *Problems of Development in the Thai Political Setting*. Paper no. 14. Bangkok: Thammasat University, Thai Khadi Research Institute.

Scott, J., and B. Kerkvliet, eds. (1986). *Everyday Forms of Peasant Resistance in Southeast Asia*. London: Frank Cass.

Sharp, L., and L. Hanks (1978). *Bang Chan: Social History of a Rural Community in Thailand*. Ithaca and London: Cornell University Press.

Suehiro, A. (1985). *Capitalist Accumulation and Industrial Development in Thailand.* Bangkok: Chulalongkorn University Social Research Institute.

Tanabe, S. (1984). "Ideological Practice in Peasant Rebellions: Siam at the Turn of the Twentieth Century." In A. Turton and S. Tanabe, eds., *History and Peasant Consciousness in South East Asia.* Senri Ethnological Studies 13:75–110. Osaka: National Museum of Ethnology.

Thak Chaloemtiarana (1979). *Thailand: The Politics of Despotic Paternalism.* Bangkok: Social Science Association of Thailand.

Therborn, G. (1980). *The Ideology of Power and the Power of Ideology.* London: Verso Editions.

——— (1984). "The New Questions of Subjectivity." *New Left Review* 143:97–107.

Thompson, E. (1978). *The Poverty of Theory and Other Essays.* London: Merlin Press.

Turton, A. (1976). "Northern Thai Peasant Society: Twentieth Century Transformations in Political and Jural Structures." *Journal of Peasant Studies* 3:267–298.

——— (1982). "Poverty, Reform and Class Struggle in Rural Thailand." In S. Jones, P. C. Joshi, and M. Murmis, eds., *Rural Poverty and Agrarian Reform.* New Delhi: ENDA and Allied Publishers.

——— (1984). "Limits of Ideological Domination and the Formation of Social Consciousness." In A. Turton and S. Tanabe, eds., *History and Peasant Consciousness in South East Asia.* Senri Ethnological Studies 13:19–73. Osaka: National Museum of Ethnology.

——— (1986). "Patrolling the Middle-ground: Methodological Perspectives on 'Everyday Peasant Resistance.'" In J. Scott and B. Kerkvliet, eds., *Everyday Forms of Peasant Resistance in South-east Asia.* London: Frank Cass. (And in *Journal of Peasant Studies* 13:36–48.)

——— (1987). *Production, Power and Participation in Rural Thailand: Experiences of Poor Farmers' Groups.* Geneva: United Nations Research Institute for Social Development.

Witayakorn Chiengkul (1982). "The Effects of Capitalist Penetration in the Transformation of the Agrarian Structure in the Central Region of Thailand (1960–1980). Master of development thesis, Institute of Social Studies, The Hague.

World Bank (1978). *Thailand: Selected Issues in Rural Development.* Background Working Paper no. 4 for *Thailand: Towards a Development Strategy of Full Participation.* Washington, D.C.: World Bank.

——— (1983). *Thailand: Rural Growth and Employment.* Washington D.C.: World Bank.

Chapter Five

Conflicts over the Deployment and Control of Labor in a Northern Thai Village

ANAN GANJANAPAN

This study investigates changing labor relations and mechanisms of access to resources in a rice-growing area of northern Thailand, following the introduction of triple-cropping and capital-intensive commercial production in the late 1970s. Despite a clear tendency for a few larger landowners to become capitalist farmers and for others, through loss of land and eviction from tenancies, to become entirely dependent on wage labor, a complex variety of tied-labor arrangements have emerged. The emergent class of capitalist farmers is variously constrained and enabled in their increasingly successful efforts to control the labor of tenants and wage laborers.

These constraining and enabling factors include only partially developed markets for land, labor, inputs, and products, despite government policies designed to accelerate the flow of capital and new technology into rural areas. They also include the efforts of some poorer villages to hang onto small plots of land for the subsistence production of the locally consumed glutinous rice, using as much family and exchange labor as possible, and of others to diversify into nonagricultural employment. Broadly political contexts are of central importance: on the one hand are the ways in which wealthy and dominant villagers are supported by state policies and institutions; on the other hand are the struggles and resistance of poorer farmers and laborers to secure the conditions of their subsistence and livelihood. These latter struggles include the large-scale mobilizations of tenants and landless villagers in Chiang Mai province and nationally in the mid-1970s on issues of land and land rent, as well as more recent local struggles over wages. This paper is based on a historical study by the author of the development of commercialized agriculture in northern Thailand from 1900 to 1981 and an intensive anthropological study of Ban (village) San Pong (population 674 in 1980), Ban Kat subdistrict, San Pa Tong district, Chiang Mai province, which is situated some 30 km southwest of Chiang Mai city (Anan 1984).

Problems of Tenancy, Landlessness, and Subsistence in Chiang Mai Province

Following the introduction of intensive commercial production of rice in the north of Thailand in the 1960s, paddy land was greatly expanded, from 400,000 ha in 1963 to 592,000 ha in 1978. In the case of Chiang Mai, already extensively cultivated, the increase was moderate, from 100,741 ha in 1963 to 116,708 ha in 1978 (*1963 Agricultural Census; 1978 Agricultural Census*; Thailand, Ministry of Agriculture 1978:16–19). This expansion of riceland, however, was not sufficient to overcome growing problems of tenancy and landlessness that developed during the 1960s. These problems were aggravated in the 1970s and early 1980s by growing contradictions in intensive commercial production that turned many bankrupt small landowners into tenants. To some extent the earlier high rate of population growth, even though it declined in the 1970s, exacerbated conditions in Chiang Mai.

Neither the magnitude of these problems nor the mechanisms by which they came about can be adequately discerned from conventional indices or from government statistics of types of land tenure. For example, the Agricultural Land Reform Office estimated that 24.54 percent of households in the north in 1976 were tenants and part-tenants (not distinguished); the percentages were 39.42 and 30.8 respectively for Chiang Mai and neighboring Lamphun province where riceland was best irrigated and most productive (Thailand, Agricultural Land Reform Office, 1977:6–10). In Ban San Pong in 1980, some 35.9 percent of cultivating households were tenants (13.7) and part-tenants (22.2), a slight decline from 1970 largely, as we shall see, as the result of evictions.

However, such figures exclude rather large numbers of landless laborers, who influence tenancy problems in the sense that they increase competition for rented riceland and hence enable landowners to demand higher rents. Several small-scale studies show that during the 1970s and in 1980 several areas of Chiang Mai had high percentages of households of landless laborers: 20.6 percent in Doi Saket district in 1980, 34 percent in a village of Mae Rim district in 1975, and 39.6 percent in Ban Kat subdistrict (San Pa Tong district) in 1978 (NESDB 1980:35; Tanabe 1981:231; Cohen 1981:195). A survey by the Northern Thai Student Center in 1975 of Mae Taeng district showed 55 percent of households landless and 35 percent tenants (*Chaturat*, 19 August 1975).

Average size of riceland holding in Chiang Mai declined from 1.2 ha in 1963 to an estimated 0.9 ha in the period 1973–1977 (Tanabe 1981:109). In Ban San Pong the average holding in 1980 was 1.1 ha. Despite historically high rice yields in Chiang Mai, despite an increase in average yield from 2600 kg/ha in the 1960s to about 3125 kg/ha in the period 1974–1980, and despite lowest average size of farm households in the country—4.3 persons in 1980 (Thailand 1983)—a large number of households faced a crisis of subsistence. We estimate that an average-sized family with average yields requires about 0.6 ha of irrigated riceland to meet minimum subsistence needs, or 1.2 ha in the case of tenant households paying half their crop in rent. The average riceland holding of tenants in Chiang Mai in 1978 was 1.1 ha. In Ban San Pong where the average paddy yield is remarkably high—4656 kg/ha in the 1980–1981 main crop—an average-sized family of 4.1 persons (1980) needed at least 0.4 ha (0.8 ha for tenants) to meet basic subsistence needs. Even here 22 percent of cultivating households (29/131) could not produce enough for their own requirements (see Anan 1984:table 6.2).

In this situation high rents became a major issue, and in July 1974, for the first time in Chiang Mai, over 100 tenants who rented about 160 ha from ten landlords protested at Mae Taeng district office (*Bangkok Post*, 30 July 1974). Early in November 1974 more than 100 tenant representatives from many areas of Chiang Mai and Lamphun provinces organized a demonstration in Chiang Mai to petition the Prime Minister. Under prevailing forms of sharecropping, tenants normally met all labor and input costs (including buffalo for plowing) and received half the gross crop. They claimed that landlords were now demanding more than half the crop (*Bangkok Post*, 7 November 1974).

Organized Resistance by Poor Farmers: 1974–76

These mobilizations took place early in the period of relatively greater political freedom from October 1973 to October 1976. They were encouraged initially by students who were funded by the government in early 1974 to go into the countryside as part of a 'Propagation of Democracy' program. The students in turn were radicalized by their experiences and became increasingly involved in agrarian conflicts and in the mobilization and organization of peasants especially in the early stages (cf. Dennis 1982:48–52). In the course of these struggles, however, peasants were able to develop their own leadership and transform the movement into one fully their own. When the government made no serious response to their demands, peasants from the north joined with others from all regions in Bangkok. Many simultaneously rallied in Chiang Mai. During this seventeen-day-long demonstration the first nationwide organization of peasants was established, the Peasants Federation of Thailand (*sahaphan chao rai chao na haeng phrathet thai*—PFT). By the end of the month the government agreed to many of their demands, in particular passing the Land Rent Control Act on 18 December 1974. The act guaranteed, in brief, tenancy contracts for six years; required one rent payment annually regardless of number of crops grown; and fixed the landlord's share to a maximum of half the crop (and less in case of poor harvest or failure) *after* one-third had been deducted by whoever paid production costs other than labor. To settle disputes, district supervisory committees were to be set up, composed of district and subdistrict officials and five tenant and three landlord representatives from each subdistrict. The PFT grew rapidly, holding meetings to inform villagers of their new legal rights, confronting landlords, and in some cases having recourse to the courts. The PFT also campaigned against official corruption—notably in the handling of subdistrict development funds (see below)—and for the allocation of land for the landless.

There was tenacious opposition from landlords and local officials from the start, and it was later estimated that the law was implemented for less than 10 percent of riceland in Chiang Mai and Lamphun, mostly owned by large landlords close to the cities (*Thai Nikorn*, 30 December 1977). From March 1975, after the formation of a new conservative coalition government, more systematic opposition to the PFT began and peasant demands for implementation of the land rent control law were rejected. Charged with subversion and labeled as communist by government controlled media, the PFT broke off its dialogue with the government and announced that it would concentrate its struggle in the villages. A period of violence began in which twenty-one PFT leaders were assassinated at the instigation of landlords and right-wing groups. Eleven of those killed were from the north including two from San Pa Tong district.

In Ban Kat subdistrict many peasants were involved in another autonomous local organization named the Peoples' Federation of Just Water Users (PFJWU) which brought together users of twenty-seven local irrigation dams. Although the PFJWU's mobilization contributed to the later construction of a concrete dam with government funds, its concern with irrigation could not address the main problems of tenants and the landless. Its considerable activities had the effect also of precluding PFT organization in the area. As a result, no doubt, of the absence of PFT organization, there was a merely nominal subdistrict landlord-tenant committee and no enforcement of other provisions of the land rent control law whatsoever. There were, however, several bitter cases of landlord-tenant disputes—especially over rent for second crops—and unsuccessful attempts by tenants to seek legal redress. When the PFT and PFJWU were disbanded after the coup d'état of 6 October 1976 and peasant radicalism was more thoroughly suppressed by the government, several landlords took the opportunity to evict tenants illegally (see below). The experience of this unprecedented degree of organized opposition by tenants, poor farmers, and the landless in the Chiang Mai region would not soon be forgotten by either side. It constitutes one crucial political context of subsequent agrarian conflicts.

An Emergent Class of Capitalist Farmers

I have elsewhere analyzed the processes of differentiation and accumulation in Ban San Pong in the decades prior to the 1970s (Anan 1984). In 1980 the ownership of irrigated riceland in the village was considerably skewed towards wealthy villagers, especially an emerging class of capitalist farmers.[1] The seven households of what I here term capitalist farmers hold an average 3.8 ha compared with an average 2 ha for 'rich peasants'. Although representing only 4.4 percent of agricultural house-

TABLE 5.1. *Distribution of Rice Landholdings in Ban San Pong*

Socioeconomic class	Households		Rice land operated	
	Number	%	Area (ha)	%
Landless	19	11.9	—	—
Poor peasant	49	30.6	7.36	5.3
Middle peasant	52	32.5	37.64	27.3
Rich peasant	33	20.6	66.0	47.9
Capitalist farmer	7	4.4	26.92	19.5
Total	160	100.0	137.92	100.0

holds they possess 19.5 percent of irrigated riceland; rich peasants, constituting 20.6 percent of households, possess 47.9 percent (Table 5.1).

These capitalist farmers are among a select number of wealthy villagers who, mainly because of their control of a large area of irrigated riceland—but also their broadly political local connections and powers—benefited most from the intensive commercial production of the early 1970s and from government policy. After 1975 the state, in addition to outright suppression of peasant radicalism, began to adopt World Bank prescriptions for a rural policy ostensibly to help the 'rural poor' (cf. Feder 1976:349–352; World Bank 1980). Instead of embarking on the implementation of land reform, however, which might have alleviated problems of tenancy and landlessness, the Thai government concentrated on the transfer of more capital and technology into rural areas. The policy aimed at keeping peasants on their farms as sources of cheap labor for the production of low-cost food and other agricultural commodities, serving further to reinforce existing rural contradictions and increasing the integration of peasants, including those with no surplus to sell, into the labor market.

Three principal policy instruments, all dating from 1975, were Local Development Schemes (*ngoen phan*), subsidized agricultural credit programs, and rice support schemes. Local development funds were intended primarily to provide wages for off-season employment on local projects. In 1975 and 1976 the Ban San Pong headman and his kinfolk kept all the wages and compelled villagers to contribute free labor. In 1977 and 1978 wages were paid only to kinfolk and clients, and in 1979 a rival elite faction removed him from office.

Agricultural credit from government-sponsored cooperatives had been available to a few wealthy villagers in the 1960s and had been used to purchase land from Chiang Mai aristocrats (see Anan 1984). From 1975 the Bank for Agriculture and Agricultural Cooperatives (BAAC) began to provide subsidized credit at 12 percent for three-year loans. This was not available to full tenants or small owners without collateral. Some small farmer debtors who were unable to repay because of crop failure lost their land; others borrowed at 60 percent interest per annum from local moneylenders—who sometimes loaned at 60 percent what they had obtained at 12 percent. Other uses of institutional credit included 240,000 baht each by three large landowners for pig farms on 0.16 ha of riceland each; 400,000 baht each for five tractors bought by two large landowners; and three six-wheel trucks, which are used mainly for the transport of rice.

Various rice support schemes have, over the years, benefited only a very few farmers, especially those who have no rent to pay, have storage facilities, and can therefore benefit from delaying sale after harvest, or those producing on a scale enabling them to benefit from bulk selling prior to harvest and from growing nonsubsistence, higher-priced, non-glutinous rice.

In 1980–1981 when paddy prices were very high, 38.7 percent of main-season rice and 99 percent of off-season rice entered the market, or 59.8 percent of total village paddy production; but this was from only 50 percent of households. Moreover 65 percent of those households which did sell rice sold to only four dealers in the village, all of them rich landowners.

In response to the agrarian conflicts of 1974–1976 mentioned earlier, several large Ban San Pong landowners transferred ownership of their riceland to their offspring by registering the land in the names of the latter. About 24 ha of riceland was transferred in this way between 1976 and 1981. In most cases parents continued to receive nominal payments from their children *(kha hua)* which, though sometimes a token amount, can be as much as half the main-season crop. These transmissions when both parents are alive are directly contrary to past practice, in which all children can inherit land (usually equally) only after the death of both parents. It is possible that large landowners adopted this strategy to avoid landlord-tenant conflicts and complications arising from the land rent control law and to avoid renting land to nonfilial tenants. The offspring of these large landowners were allowed a free hand to cultivate their parent's riceland with the help of hired laborers. This development also indicates an attempt by parents to alleviate dissension among their children, a phenomenon which has increased considerably as a result of the growing profits from intensive commercial production.

Both agrarian conflicts and intensive cultivation are also major factors influencing the nature of land sales in the period 1973–1981. After experiencing a period of intense conflict with tenants at that time, some Chinese traders in Ban Kat who owned land in Ban San Pong felt that income from renting their land was not worth the risk. After 1976 they began to sell all their 2.88 ha of riceland in Ban San Pong to wealthy villagers. Unlike large landowners in Ban San Pong the traders could not opt for taking over the cultivation themselves, even with the help of hired laborers or shared-cost leasing arrangements (see below), because they were preoccupied with other more remunerative enterprises. By taking advantage of the high price of riceland, most traders reckoned that they could earn more from interest on bank deposits than from rent.

Land prices have been driven up because of the higher returns from triple cropping and increasing demand by wealthier farmers. In 1967 a price of 39,375 baht/ha is reported (Cohen 1981:176–177); ten years later a Ban Kat trader sold 1.28 ha to two Ban San Pong villagers at 109,375 baht/ha and in 1979 another sold for 156,250 baht/ha or US$7812. Indebted small landowners have benefited in the short run. They can at least hold on a little longer, selling small portions of land to meet debts. In this way five indebted villagers sold an average of 0.16 ha each between 1977 and 1981 compared with nine villagers selling an average of 0.38 ha each to creditors between 1973 and 1977. However, the landless and other poor villagers are now entirely excluded from the land market. Small landowners have redoubled their efforts to hang onto their holdings; but the reduced incidence of land transfers within domestic groups suggests that kin ties are playing a diminishing role in resisting the process of land concentration. In one case four poor related families inheriting 0.08 ha each decided to pool their land and cultivate it by rotation one year in four in order to avoid becoming completely landless farmworkers. Overall the concentration of riceland into the hands of the wealthy villagers continued its previous increase, if slowly, during the period 1973–1981. The process has been assisted by the availability of credit from the BAAC and commercial banks which were set up in the district from 1972. From 1977 to 1981 there were transfers of land by sale of 7.4 ha (sixteen cases). On the other hand the extent of land mortgaged began to exceed that of land sold. Between 1977 and 1981, some 11 ha (fifteen cases) was mortgaged: 7.6 ha to the BAAC, 3.4 ha to the Bangkok Bank and Thai Farmers Bank.

Unlike the government's financial institutions, the commercial banks encouraged villagers to purchase large tractors and trucks, rather than buy additional land. In this way the banks can benefit not only from the interest on the loans but also from the profit from sales of tractors, trucks, and other commercial transactions in which they have a direct financial interest. The loan policy of commercial banks also began a process of advancing their control over riceland, which is very likely to contribute to the real transfer of riceland ownership from the villagers to the bank and to have significant repercussions on the agrarian structure and agrarian production in the future.

The capitalist farmers are investing heavily and diversifying into off-farm and on-farm enterprises: trading, machinery hire, haulage, pig raising, and moneylending. Although they are in a better position than others to operate their own riceland using wage labor, they also continue to rent out the largest area of riceland, averaging 2.24 ha per household, compared with 0.6 ha for rich peasant households, and a much higher proportion—52 percent as compared with 15.4 percent—to 'nonfilial' tenants (see below). Capitalist farmers continue to rent out land not only because they are engaged in other enterprises, but also because of an uncertain labor supply as a result of the struggles of farm laborers. These strategies of resistance by labor and response by landlords in devising new and transformed means of control of wage labor and tenant labor are discussed fully below. Recently increasing numbers of poor and middle peasants have been forced to rent out their land too—mostly, however, because of indebtedness as a result of failure in intensive commercial production or inability to produce second and third crops for which high capital inputs are required. It is in off-season cultivation that the rental market is most developed. To explain this, some preliminary discussion of the intensive use of irrigated riceland and complex cropping patterns is required.

Triple-Cropping

In 1978 Ban San Pong villagers were triple-cropping on their irrigated riceland, a practice a few households had begun experimentally ten years earlier. By the 1980–1981 season there was triple-cropping on nearly 40 percent (55.04 ha) of all irrigated land (138.96 ha) cultivated for main-season rice; a further 5.52 ha was being rented to tenants from other villages. Almost all of the remaining irrigated riceland is double-cropped. The main-season rice crop is about 92 percent glutinous rice, a subsistence crop for the majority of producers. The second crop on most land is soybeans, planted on about 87 percent (122.36 ha) of irrigated riceland; of the remainder, about half is planted to a second rice crop and about half to garlic, onions, and chili, or left fallow.

The third crop is entirely planted to nonglutinous rice for the market: 90 percent to the Thai Rice Department's KK7 variety, 10 percent to KK1. These non-photoperiod-sensitive, high-yielding varieties, which can be planted in any season, were first introduced into the area in the early 1970s. They were not adopted on any scale until the late 1970s when more capital was flowing into rural areas and more cultivators could afford the necessary chemical fertilizers, herbicides, and pesticides. These contributed to a rapid increase in yield from 3125 kg/ha in the early 1970s to 5500 kg/ha for main-season rice in 1981.

Improved water supply also facilitates triple-cropping. In 1981 thirteen new tubewells, capable of supplying 3 to 10 *rai* (0.4–1.6 ha) each, were added to the existing eight wells. The local irrigation system was also improved with local development funds between 1975 and 1978. Most important has been the introduction of power tillers, which increased in number from twelve in 1978 (Cohen 1981:111) to twenty-six in 1980. These machines, manufactured in Thailand with imported Japanese engines, cost about 16,000 baht or the price of two full-grown buffalo. They allow cultivators to prepare land in at least two weeks less time than the use of draft animals requires, and they can be used after the first rainfall when the ground is still too hard for plowing with buffalo. This is crucial for the triple-cropping system because of its extremely tight schedule, with some crop seasons overlapping.

This triple-cropping system has led to a complex development of tenurial arrangements which will be discussed in the following section. The development provides opportunities for landless villagers to rent some riceland for particular seasons, but in the long run it may jeopardize the well-being of poor villagers because the third rice crop is so highly capital intensive that most poor tenants cannot afford to undertake it. In this case, as we shall see, landowners provide capital and take more control of the management of cultivation in a form of 'shared-cost leasing' which will gradually force tenants into the role of wage laborers facing an uncertain, subsistence life. Small indebted landowners may benefit, however, from being able to rent out land for one or two cropping seasons without endangering their main-season subsistence rice crop.

Changing Forms of Tenancy

Prior to the 1970s the principal form of tenancy which had emerged was the renting of land by independent households of children from their parents, which I here term 'filial tenancy.' The percentage of riceland rented in this way has declined from 70.3 percent in 1970 to 57.3 in 1980 and has assumed a more commercial character. This has had a more marked effect on poor tenants (renting an average 0.67 ha) for whom the proportion of land rented from parents has declined from 80 percent in 1970 to 45.5 percent in 1980. The proportion rented by middle peasants (average area 1.09 ha) has remained at about 50 percent, while rich peasant households (average area 1.62 ha) rent *only* from parents (see Anan 1984). Although for the main rice crop filial tenants pay half the crop in rent, they have greater security (and some anticipation of inheritance) and are likely to have more assistance with labor and other inputs than 'nonfilial' sharecroppers. Children of rich peasants and capitalist farmers may pay only a nominal sum (*kha hua*) to their parents, sometimes at their own discretion. In one case a rich filial tenant paid cash rent in advance (*lang na*, see below): 5400 baht/ha per annum, compared with a more usual 8750 baht/ha per annum for nonfilial tenants.

For the second (soybean) crop, filial tenants pay no rent; though some are able to rent only for the second crop, in which case they pay a quarter of the cash rent (500 baht) paid by nonfilial tenants. Only rich peasants are likely to allow their offspring to cultivate the third (rice) crop for no rent (except perhaps for a discretionary *kha hua*). Interestingly, access to land for the cultivation of second and third crops may be extended to offspring still living within the parental household. Thus a household which operates as a single unit in the main rice season separates financially in the cultivation of cash crops. In this way intensive commercial production of cash crops allows children to prepare for separation from the parental household earlier than in the past and may lead to increasingly commercial relationships between parents and offspring. It also poses problems for approaching households as units of production except in main-season cultivation.

Even with triple-cropping many Ban San Pong villagers found no access to their parents' riceland. Poor peasants have to look for riceland elsewhere. In 1980 a number of poor and middle peasants were able to transform small upland holdings, in an area known as Huai Manao, into rain-fed riceland. This process usually required large amounts of capital for hiring tractors to level the land, which only a few poor villagers could afford. After 1974 when the government intervened with a plan to allocate land at Huai Manao, access to this area became even more limited because of official corruption (Cohen 1981:267–268). Many villagers were thus forced to become nonfilial tenants on the land of others.

The percentage of riceland rented from other than parents (whether kin-related or not) by poor peasants has more than doubled from 21.5 percent in 1970 to 54.2 percent in 1980–1981; the proportion rented by middle peasants has remained about the same. The most common form of rent (on 74.5 percent of rented land) was half the crop yield with the tenant bearing all input costs. Some landowners who were also capitalist farmers demanded even higher rents: in two cases (5.6 percent of rented land) they required two-thirds of the crop; and in two other cases (11.3 percent of rented land) they demanded an additional advance cash deposit or premium (*wang ngoen*) of about 6250 baht/ha—on which interest is paid at 60 percent per annum—returnable on termination of tenancy.

A form of fixed cash rental (*lang na*) has become a viable option for landowners not resident in the villages where their tenants live—and thus not having to supervise cultivation—and those who need large sums of cash quickly. There were three cases of landowners in other villages renting *lang na* to Ban San Pong villagers, three cases of Ban San Pong landowners renting to tenants in other villages, and two cases of Ban San Pong landowners renting *lang na* to tenants in their own village. Of the latter, one was a rich peasant needing to pay off a bank loan and the other a poor peasant owing to a local moneylender.

Most land rented in this way is available for triple-cropping, though there were several cases of renting *lang na* for a single crop season, especially renting out by small landowners. Rents are lower the longer the term of the tenancy. In one case, 0.4 ha was rented for three years at 10,000 baht/ha per annum; in another, 0.64 ha rented for five years at 7500 baht/ha per annum. Small landowners in urgent need of cash might demand as little as 6250 baht/ha per annum.

I have elsewhere attempted to construct an annual production account for an average farming unit of one *rai* that can support triple-cropping, in order to compare the costs and benefits in production as between different cultivating units under various forms of tenure (Anan 1984). Here I present only a summary of net incomes. Landowners made approximately the same amount from rent from nonfilial tenants as if they had cultivated themselves, except when contracting a fixed-term tenancy with cash in advance (*lang na*). Filial tenants paying only half of the main-season crop as rent for the whole year have a substantial return. Nonfilial tenants renting for a period of years on a *lang na* basis show a very modest return. But nonfilial tenants paying half-shares for all crops barely break even; and those paying two-thirds for the first crop and half for the other two, and those additionally paying a cash deposit or premium (*wang ngoen*), operate at a considerable loss when their labor is valued in market terms.

Nonfilial tenants are able to cultivate because they substitute their own labor for most marketable inputs. These tenants rarely hire labor, and those without draft animals had to contribute household labor to labor exchange networks in order to gain access to power tillers. For off-season rice they cannot avoid hiring power-threshers, because the paddy is reaped when still mostly green and cannot be threshed by hand. For other items such as seed, tenants try to produce for themselves rather than buy it on the market, which is more difficult in the case of soybean seed for those who have no upland farms. These nonfilial tenants even use their own labor for weeding instead of purchasing the herbicides used in soybean and off-season rice cultivation.

If tenants were to turn to work exclusively as wage laborers they would face great uncertainty in their subsistence, even though they would sometimes be able to buy rice at lower costs than they would have to spend producing it on their rented plots. Rice prices vary considerably, however, even within the same year. Accordingly, sharecropping provides tenants with some security in the face of seasonally fluctuating labor and product markets.

From the landlord's point of view, this variety of forms of tenancy and sharecropping represents attempts to capture the largest surplus possible directly from the tenants. With some monopoly control of land, most landlords can present themselves as patrons to their tenants, assuring the latter's subsistence. The landlords in turn are able to use such patronage relationships as a means of securing their tenants' labor supply in an increasingly uncertain labor situation (see below) which makes cultivation on the basis of wage labor highly risky. At the same time, they are ensured a well-disciplined workforce that does not require direct supervision. This is an important advantage, given large landowners' extensive involvement in other activities.

Another reason why sharecropping persists is that a fully developed cash rental market for land does not yet exist. Such a market began to emerge with the development of tenurial arrangements on the basis of fixed cash rents (*lang na*), which through market mechanisms make agricultural production profitable. But this practice requires access to capital and credit, and as long as these markets remain underdeveloped this form of tenancy cannot easily lead to a fully developed rental market. Such tenurial arrangements have been tried mainly by small landowners whose inferior position and need for cash leave them unable to bargain for high rents. Thus a fixed cash rent is not yet a viable alternative either for the large landowners who cannot demand high cash rents or for poor villagers who have limited access to credit.

Although sharecropping arrangements are in many ways functional to both landlords and tenants, they also give rise to tension and antagonism. Generally landlords are careful not to demand more than half of the harvest from main-season paddy as rent, because this rice is so essential for sustaining the life of their tenants. Instead landlords may try to increase rent on other cash crops. However, in 1977 one landlord began to impose a rent of two-thirds of the main-season rice crop on a new tenant. A year later the tenant complained that he could not make ends meet and asked the landlord to revert to the previous rent of half the crop. The landlord agreed, but on the condition that the tenant paid a premium in addition to the half-share (Cohen 1981:193). Two tenants, however, were able to pay two-thirds of their main-season rice crop as rent in 1980–1981 because the land they rented was sufficiently productive to maintain their small families.

Under these conditions, many tenants in 1980–1981 thought that the rice they produced from their rented land was more expensive than rice currently available in the market. Although no tenants yet wanted to leave the land they rented, they voiced their frustration, expressing the opinion that they might be better off working as wage laborers and buying their rice from the market. This effort to squeeze tenants underlies the growing antagonism between landlords and tenants in recent years.

Despite this antagonism, tenants still require secure employment and landlords still need to ensure their labor force. But as agricultural production has come to require more and more new capital in order to increase productivity, landowners have begun to score heavily in their struggle to control labor. Without capital, tenants cannot by themselves cultivate intensively. Some landowners who need the labor of these poor peasants turn to what I term here 'shared-cost leasing' arrangements—*yia na pha nai thun*, literally 'work the fields sharing with the owner of capital'—providing tenants with new inputs that enable them to work the land more efficiently.

The Modification of Sharecropping Arrangements

This arrangement began to develop at Ban San Pong following a wave of evictions during 1976 and 1978. In 1978 a few landowners tried to furnish some of the costs of production as a means of claiming a greater share of the crop from the new incoming tenants. In one case a new tenant received only 36 percent out of the yield from the main-season crop after paying all input costs. For the second season, however, the landowner supplied fertilizer and soybean seed but allowed the tenant to cultivate only one-third of the plot without rent while the landowner worked the remainder himself (Cohen 1981:192). In another case a new tenant received 30 percent of 10,000 kg harvest on 2.1 ha but was entitled to half of any yield above 10,000 kg. This tenant was responsible for all production costs except for chemical fertilizer. In the dry season the landowner supplied onion seed and claimed no rent for that crop.

Thus landowners not only claim more than half the main-season rice crop but also increase their control over decision-making: the kind of crop to be planted and the extent of land to be cultivated, especially in the dry season when landowners can choose to work for themselves. By imposing all labor costs on tenants, landowners can be relieved of the uncertainty and burden of labor management. This in turn allows them to diversify and assume more entrepreneurial roles (in trading, transport, contract pig farming, machinery hire, etc.). Tenants are increasingly forced into the role of permanent farmworkers, and this is clearly how it is perceived by landowners, who insist that they did not rent their land but 'hired' (*chang*) their 'tenants.'

Tenants under these arrangements tended to be paid in fixed amounts of the main-season rice crop, rather than in a percentage of the yield. This allowed the landowners to gain all the benefits from the potential yield increases as a result of the application of new inputs, which they contributed. For tenants this meant a lower return on their labor, as they got less from these fixed amounts than sharecroppers usually get from half shares: in some cases close to a quarter of the yield. Tenants complained that they had to cultivate at a loss. They had to spend roughly the same amount of cash on wage labor or on midday meals served to exchange labor, as the price of the paddy they received as their share. This meant that they got close to nothing for their own labor. Most cannot continue to work for more than one season.

In response, large landowners (both rich peasants and capitalist farmers) tried to modify their relationship with contracted workers (now a more appropriate term than 'tenant') within the general form I have termed 'shared-cost leasing.' They tended to provide all the capital needed for cultivation, which turned them into *nai thun* (capital-owners) in the eyes of the workers. In 1980–1981 three landowners supplied all capital inputs for seed, chemical fertilizers, gasoline, and power tillers. The contracted cultivators undertake to provide their labor and to engage in exchange networks in place of the owners—becoming more like permanent laborers on the plots—and received only a small fixed proportion of the yield. One of them got only 200 *thang* out of more than 800 *thang* yield from a 10-*rai* plot; he was, however, allowed to cultivate off-season rice with capital provided by the landowner, including a water pumpset, enabling him to cultivate land that would otherwise have been left fallow. In such cases landowners benefit considerably by deducting all costs of production before dividing the yield equally with the worker. The worker in this case felt morally indebted to the landowner for allowing him to grow the off-season rice, which is highly capital intensive and risky.

Thus 'shared-cost leasing' arrangements amount to a combination of tenancy and wage employment. Although such arrangements are still the minority of cases, their development indicates that more landless villagers are being forced into the ranks of wage earners (*khon hap chang*) who are no longer able to maintain themselves as tenants.

Agricultural Wage Labor

A growing number of smallholders have been forced to become dependent rural workers, with greater insecurity of subsistence, because they have lost their own land or have been evicted. Large landowners have to rely more on the employment of hired laborers who are increasingly unattached and mobile. Conflicts in relations between employer and laborer have currently become more evident—if more subtle than those between landlord and tenant which resulted in the overt violence of the mid-1970s—and are increasingly coming to center on the issue of wages. Employers and workers are fully conscious of their respective struggles to lower or raise wages. But on the one hand employers constantly express concern over 'labor shortage' and high wages at times of peak labor demand; and on the other hand workers frequently express the fear that they will not be rehired (*kua poen bo chang*). This latter indicates the various pressures and threats imposed by employers in their efforts to reduce their wage costs and also to undermine workers' resistance.

The demographic context in Chiang Mai province is one of a declining rate of population growth, with a real increase in population growth from 1970 to 1980 only about half that of the preceding decade. Average household size has also fallen from 5 members in 1970 to 4.3 in 1980, and in Ban San Pong it is only 4.1. With declining household size, problems of insufficient labor supply faced many families in Ban San Pong, especially the poor and the capitalist farmers whose households contain an average 2.5 and 2.8 potential adult workers respectively. There has been considerable net out-migration from the village since 1970 with increasing numbers leaving for land frontier regions, the upland area at Huai Manao, Chiang Mai city, or to work or study in Bangkok. Several women from poor households have left to work as prostitutes in Bangkok, while many children of rich peasants and capitalist farmers have gone there for university education. Few of these are likely to return. For the first time there is now a small number of households which are entirely nonagricultural: engaged in housebuilding, woodcarving, small shopkeeping, etc. Many more who are engaged more than casually in off-farm work rarely return to agriculture even at harvest time. These include about thirty mainly young men engaged in woodcarving, which is subcontracted by four village craftsmen. Teenage girls are involved in casual off-farm work: the poor are engaged largely in low-paid bamboo weaving for hats; those who can afford a 2400-baht knitting machine and tuition fees can earn money knitting sweaters at a little more than the agricultural wage. This too is subcontracted by rich villagers from city shops which take advantage of cheap rural labor. Many poor women turn increasingly to petty trading, selling vegetables and prepared food in local markets.

There is now a growing demand for women wage laborers from poor and middle peasant households who would previously have been engaged exclusively in domestic work and exchange networks. In addition, some laborers from outside the village are employed, especially in August when competition for labor for harvesting off-season rice and transplanting main-season rice is most intense. Forty-five laborers from other villages were hired each day for the two rice crops in 1980 and 1981. The average number of laborers needed each day during the peak August season is 270, of which 225 Ban San Pong villagers worked daily on average during this period: 120 wage laborers, 65 in exchange networks, and 40 members of the landowning households. In 1980–1981 most households did not have sufficient labor for their own holdings. An increasing number of poor and middle peasants also found themselves busy working as wage laborers and in exchange networks, especially in transplanting main-season rice, and so could not afford to spend a great deal of time on their own holdings.

The use of exchange labor (*ao mu ao wan*, literally 'take turns, take days') for rice cultivation has decreased at an accelerated rate since the introduction of triple-cropping. For main-season rice cultivation in 1980, landowners used exchange labor for transplanting about 30 percent of total area, a little more for reaping, and about 50 percent for bundling and threshing. Capitalist farmers used it for reaping and threshing only and not at all for transplanting. The lower down the economic scale, the greater the dependence on exchange labor. Yet precisely the poorer are more likely to be involved in harvesting off-season rice on their own fields or to be working as wage laborers for rich farmers. Moreover since not everyone cultivated a third crop, it was often difficult to form a sufficiently large exchange labor group. Rich and middle peasants are more likely to hire wage labor which they reckon can be cheaper than participating in exchange networks given the increased costs of providing a midday meal. Delays in cultivation due to insufficiently large exchange groups can be costly in intensive commercial production of the off-season rice crop. In 1981 under 10 percent of holdings in owner-cultivated fields employed exchange labor, even less for tenant-cultivated fields.

The withdrawal of many cultivators from exchange networks may at first lead to a reduction in wages, as more become wage laborers. However, as the number of exchange networks becomes even fewer and the labor of many rich and middle peasants is no longer available, then upward pressure on wages will develop. In an attempt by larger landowners to keep down wages and to secure labor, a compromise between hired and exchange labor has emerged, especially for harvesting main-season rice. This is a task contract hiring that is paid in kind as a percentage of harvest, known as *chang roi thon*, in which the hired laborer participates in networks of exchange labor on behalf of the employer, thereby recruiting others to work with him or her. Just as 'shared-cost leasing' could be seen as developing into a partly disguised form of labor hire, so *chang roi thon* might be considered a disguised form of tenancy. The landowner secures labor without recourse to capital for cash wages but assumes none of the responsibilities of a tenurial relationship.

In contrast to sharecropping, *chang roi thon* can be used for specific tasks, mainly for harvesting (on 40 percent of owner- and tenant-cultivated fields in the 1980–1981 main season) and a combination of transplanting and harvesting. By plowing faster and more cheaply with power tillers, landowners are now able to replace the sharecropper's obligation to prepare the land. Rich landowners and rich tenants relied most on this form of hiring, preferring it to renting out all their land to sharecroppers. They paid only about 13–15 percent for harvesting or 18–20 percent for a combination of transplanting and harvesting, compared with 33–50 percent under customary sharecropping practices. Some landowners resorted to outright deceit to lower their wage bill still further, by substituting low-priced glutinous rice for the higher-priced nonglutinous crop harvested. When workers discovered they had not received what they had harvested, employers retorted that they did not eat nonglutinous rice anyway.

Some poor and middle peasants whose household labor is insufficient to allow them to participate in exchange networks turned to hiring on the *chang roi thon* basis since their main-season crop is for consumption, not sale, and they have little available cash to pay wages. Although workers benefit from this form of hire to the extent that they secure some access to subsistence rice, in 1980–1981 some began to voice discontent over the low percentage of harvest received. Most had a clear understanding that they received a wage in kind which was below the going market wage. They argued that since 13–15 percent of harvest covered only the cost of serving the midday meal to the exchange helpers they recruited, working on the basis of *chang roi thon* was little better than buying rice at the market price; and if the value of their own labor was included in the calculation, it was like buying rice at more than the market price. Many turned more to accepting daily cash wages, but most felt a lack of bargaining power, especially during the harvest of main-season rice, when there are few other cultivating tasks to compete for labor. They expressed the fear that employers would retaliate by not rehiring them.

Other forms of daily wage payment in kind exist: marginally in the case of *choi ao khao* (helping in getting paddy) in which poorer villagers join in harvesting with the intention of asking for rice; more frequently in the hiring of laborers for plowing with draft animals. But intensive commercial production with triple-cropping has dramatically increased the monetization of wage-labor employment. For example, laborers who own power tillers rarely accept wages in kind. Cash wage labor is used extensively—on 90 percent of the holdings of the capitalist farmers—for transplanting main-season rice when the cropping schedule is so tight, but much less for harvesting main-season rice when there is no conflict of schedules and less competition to market glutinous rice. The highly commercial soybean and third nonglutinous rice crop are highly dependent on cash wage labor, however, especially on larger holdings.

During 1980–1981 most laborers were paid on a daily basis at 20 baht per day, raised to 25 baht per day for the main-season rice harvest when the farm-gate price of rice rose to 25 baht per *thang*. In Ban San Pong there is equal pay for men and women, in contrast to remoter villages where women may receive 5–10 baht per day less than men. This reflects an increasing demand for the labor of women and teenage girls, especially those from middle peasant households. For some tasks cash wages are paid on a piecework basis, including land preparation for soybeans, soybean husking, and plowing for rice cultivation. Skilled workers can earn high wages from piecework, but it is precisely these tasks which are being replaced by machines. Piece-rate contract hiring (*chang mao*) is rarely used—for example, on only 15 percent of all owner-cultivated fields for transplanting main-season rice, much less for reaping and for transplanting off-season rice. The decline in this form of labor hire again indicates intense conflict over wage-labor employment. Although *chang mao* workers can speed up their work, they can also bargain for higher wages in an individual contract in a peak period. Most employers have turned to casual laborers on a fixed daily cash wage.

The rapid introduction of new machinery is another method that employers have used to deal with labor problems. By 1980 with subsidized credit from the BAAC, thirty households were able to own and operate twenty-six power tillers. These and the five locally owned tractors were used to plow about 70 percent of main-season riceland and 98 percent of the off-season rice area. With less land available for grazing under triple-cropping, the cost of feeding draft animals has increased rapidly, reducing the opportunity for landless villagers to hire their buffalo. Power tillers are also used for husking the soybean crop by rolling over it and, with modifications, for cutting rice stalks after harvesting and for threshing paddy.

Other employment strategies have been briefly referred to earlier. These include the extension of the labor market to other villages—in some cases up to 10 kilometers distant—and the employment of women and young girls for equal pay, especially from middle peasant households who had not previously entered the labor market. To a lesser extent some rich landowners have resumed the older practice of hiring attached farm servants (*luk chang*) who work exclusively for one employer on a yearly basis and receive fixed wage payments in kind. Though small in scale this practice allows large landowners to divide the ranks of wage laborers into permanent and casual. Attached workers are less likely to join others in wage bargaining (cf. Byres 1981:438) as was evident in 1981 when most *luk chang* sided with employers against a team of casual laborers.

A number of tactics of resistance by laborers to low wages and other terms of employment have been referred to in previous sections. What is clear is that long-established forms of individual bargaining within relationships of patronage have given way to more antagonistic relations between employers and unattached and mobile labor. Moreover in contrast to the period before 1978, workers' actions in 1981 were in several cases collective in nature and in a few cases directly confrontational.

This development was in part based on workers' experience of working in the fields together as a team. Taking advantage of a serious labor shortage in August 1981, some wage laborers formed two new teams to demand an increase in daily wage from 25 to 30 baht. They threatened that they would look for alternative work outside the village. One team, just before August, joined together to plant rice on rain-fed land belonging to one member at Huai Manao, agreeing to divide the harvest equally giving a token rent to the owner who could not otherwise have made full use of it. This forced the employers to settle at 28 baht per day. Workers from another team, led by a more compromising old-style work team leader (*hua kaeng*), began to join the two new teams. Furious landowners, led by the village headman, accused the team of being communist-inspired. The headman threatened arrest, claiming that it was illegal to organize a farmers' group without government authorization, and announced publicly, and menacingly, that the team leader would not be able to stay alive longer than three years—or he would resign. By the end of the year there had been no physical violence except for minor fistfights. This intimidation caused some workers to withdraw from the team, though one older woman member publicly stated that she was no longer afraid of the headman's threats: so long as workers stayed together in the team, the large landowners could not eliminate them all.

Landowners then resorted to employing workers on a daily wage basis, but workers responded in turn by going slow, lengthening the time spent on each plot so as to create a shortage for some landowners. As a result the off-season rice of several owners was damaged. The new team of workers was helped in its struggle by the fact that about twenty households of poor and middle peasants from one village section decided to withdraw their labor from the market and return to cultivate their own small subsistence plots by mutual exchange of labor.

Conclusion

This study reveals a number of complex changes in production relations and in strategies of accumulation and subsistence, following the rather recent and rapid introduction of more capital-intensive triple-cropping and mechanization. Features of the overall context included the closing of the local land frontier, a decline in the rate of population increase, and net out-migration. A crucial political dimension included preferential state financial and legal support for rich landowners and village officials.

Advanced payments of cash rents for fixed terms had been introduced, but sharecropping predominated. The multiple forms of sharecropping arrangements were constantly, almost annually, being renegotiated, redefined, and redistributed. They varied according to type of contract, by crop and by season, by kinship status, and, decreasingly, by client status. Landowners' shares of half to two-thirds were common, and sometimes cash premiums were additionally demanded. Plowing by power tiller has reduced the status and share of some tenants to receiving wages in kind as a percentage of harvest, or harvesting only, or transplanting and harvesting. Sharecropping was increasingly on nonparental land, and parent-child tenancies were becoming more commercialized.

There had been a marked decline in exchange labor arrangements as richer households withdrew from exchange networks and the poor turned to wage labor and to intensifying household labor on small plots. Labor was becoming detached from cliental relations and locally more mobile. The labor of women and 'middle peasant' households began entering the wage-labor market on a far greater scale. There was some development of collective bargaining by agricultural workers, some of whom, especially at seasonal peaks of labor demand, could withdraw their labor to work on subsistence plots or off-farm work.

Despite a general tendency towards land accumulation and landlessness and the employment of wage labor, prevalent forms of sharecropping tenancy constitute a complex array of means by which landowners capture more of the surplus labor of tenants and acquire greater control over agricultural decision-making without taking on an increased burden of management. Many tenants have become—and certainly perceive themselves as having become—poorer as a result of producing for the market. But the importance of producing the first subsistence rice crop, and the uncertainties of wage labor, prevent sharecroppers from abandoning this means of access to land.

Parties to these production relations are keenly aware of their different interests and bargaining strengths. Resistance and conflict, sometimes bitter, have become marked features of village life as local economic struggles to control labor are increasingly linked with the exercise of power and more comprehensive forms of social control.

Notes

1. 'Socioeconomic class' in this paper is defined both in structural terms and in terms of peasant consciousness as it has developed in the specific northern Thai historical context, in which from the early twentieth century an emphasis on property relations has gradually given way to an emphasis on labor relations:

1. Landless laborers are those who live in rural communities but own no agricultural land and rely exclusively on wages from agricultural and other labor and on some income from petty trade.
2. Poor peasants are those who are unable to maintain themselves from household production—with size of holding under 2.5 *rai* (0.4 ha)—and have to rely partly on renting in land and partly on exchanging their labor power for wages on a regular basis.
3. Middle peasants are able to maintain themselves mainly from their own household labor and land—with size of holding from 3 to 6 *rai* (0.48–0.96 ha)—but for stability some may have to rent some or all of their land, without necessarily or regularly selling their labor power.
4. Rich peasants are those who own superior means of production—with size of holding from 7 to 15 *rai* (1.12–2.4 ha)—and are able to accumulate capital and appropriate surplus through investment in production, by letting out part of their land and by hiring other peasants to work for them.
5. Capitalist farmers are those able to invest heavily in both farm and off-farm enterprises.

References

Anan Ganjanapan (1984). "The Partial Commercialization of Rice Production in Northern Thailand (1900–1981)." Ph.D dissertation, Cornell University.

Cohen, Paul T. (1981). "The Politics of Economic Development in Northern Thailand, 1967–1987." Ph.D dissertation, University of London.

Dennis, John V. (1982). "The Role of the Thai Student Movement in Rural Conflict 1973–76." M.S. thesis, Cornell University.

Feder, Ernest (1976). "The New World Bank Programme for the Self-Liquidation of the Third World Peasantry." *Journal of Peasant Studies*, vol. 3, pp. 343–354.

NESDB Staff (1980). "Doi Saket" [A report on Doi Saket district]. *Warasan Sethkit lae Sangkhom*, vol. 17, pp. 33–44.

Tanabe, Shigeharu (1981). "Peasant Farming System in Thailand: A Comparative Study of Rice Cultivation and Agricultural Technology in Chiang Mai and Ayutthaya." Ph.D dissertation, University of London.

Thailand, Agricultural Land Reform Office (1977). *Kankrachai Kanthiikhrong Thidin Phua Kasettrakam Pi 2518* [Distribution of agricultural land under tenure in 1975]. Bangkok: Agricultural Land Reform Office, Ministry of Agriculture and Cooperatives. (Mimeo.)

Thailand, Ministry of Agriculture (1978). *Agricultural Statistics of Thailand: Crop Year 1977–78*. Bangkok.

Thailand, Office of Prime Minister (1964). *1963 Agricultural Census*. Bangkok.

——— (1980). *1978 Agricultural Census and Report*. Bangkok.

——— (1983). *1980 Population and Housing Census*. Bangkok: National Statistical Office.

World Bank (1980). *Thailand: Towards a Development Strategy of Full Participation*. Washington, D.C.: East Asia and Pacific Region Office.

Newspapers:

Bangkok Post (in English)

Chaturat (in Thai)

Thai Nikorn (in Thai)

Part Three

The Philippines

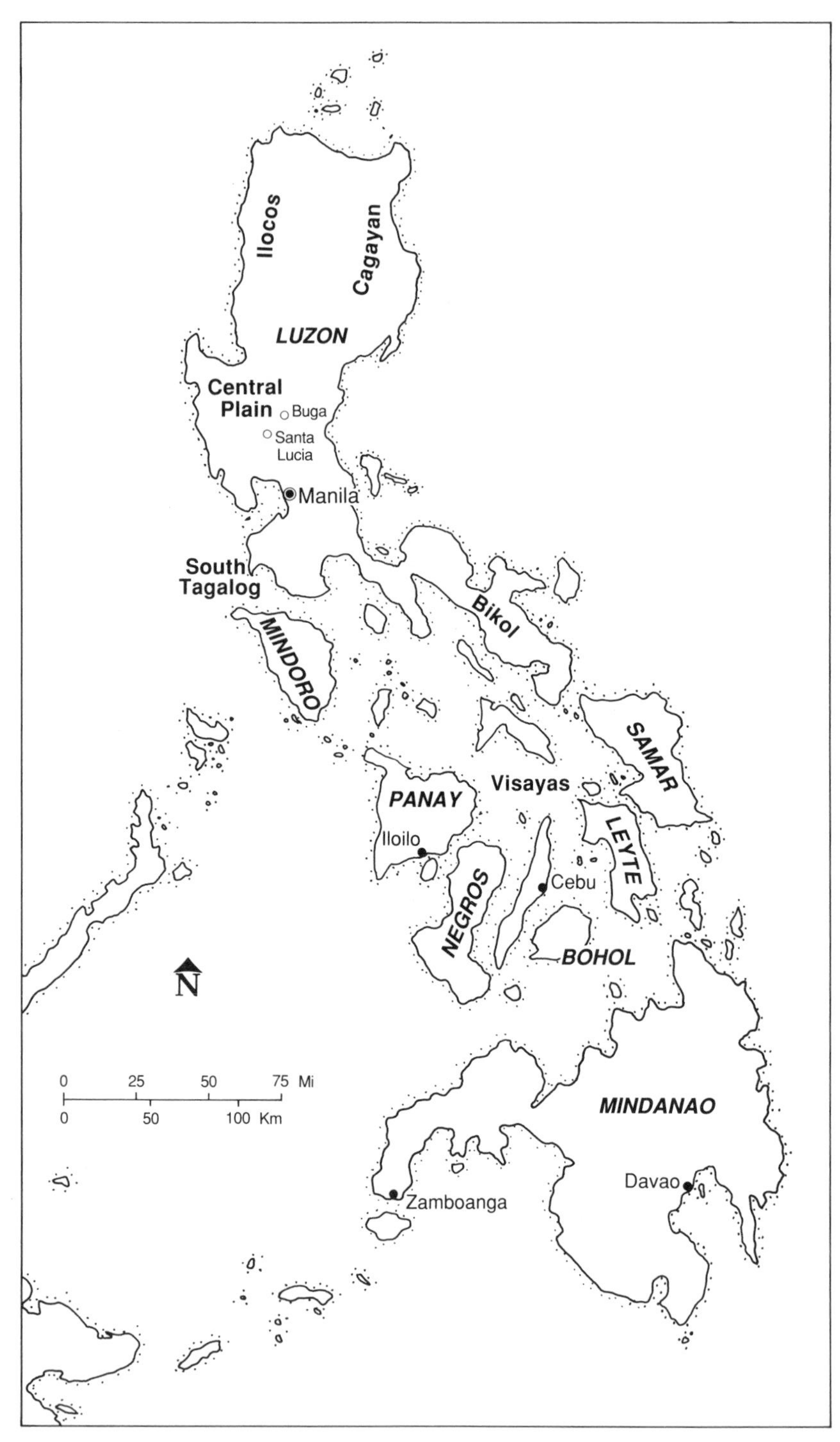

Map 2. The Philippines

Chapter Six

The Philippines: Agrarian Stagnation Under a Decaying Regime

BRIAN FEGAN

Philippines agriculture is characterized by a large number of smallholders and tenants growing rice, maize, and coconuts on small farms averaging about 2.8 ha on about 80% of the cultivated area. This pattern is typical of low-value, nonperishable crops. A growing number of large multinational or joint venture corporate wage-labor farms grow high-value export fruit in the recent frontier areas of Mindanao and Palawan islands.

The large-farm pattern is typical of crops that require scientific, high-capital production processes, and especially perishable crops that require high investment in postproduction canning or cooling and packaging. A dependent ring of closely supervised small and medium farms may surround the nucleus estate and shipping corporation. They are tied by monopoly growers' contracts to use all and only the nucleus corporation's inputs, follow its practices, and to sell all product to the nucleus at prices it dictates.

Although land, labor, and the crop have been market commodities since at least the early nineteenth century in core areas of the Philippines, the predominant forms in which product has been appropriated from the agricultural population have been by tenancy, debt, and merchant profit. According to the 1970 census, rice accounted for 34.9% of harvested area, maize 26.8%, and coconut 21.1%. All are nonperishable crops that do not depend on capital-intensive processing or fast marketing channels. Their predominant production form has been the small farm operated with a combination of family, exchange, and hired labor. Of the major crops, sugar with 4% of area has been characterized by the 'factory in the field' large farm with hired labor only on the island of Negros. Elsewhere, notably in Pampanga and Tarlac, even sugar has been a small-farm crop with 66% of farms nationally under 5 ha, many of them tenanted. Nationally, all farms average under 3 ha, rice about 2.6 ha, maize about 2.8 ha, and coconut 3.9 ha.

Value added per hectare in the export fruit sector is estimated at three to thirty times that in the three traditional crops accounting for the largest area: rice (P7750), maize (P2880), and coconuts (P3860). Recent estimates place the value added per hectare of export crops at mango P68,490, pineapple P40,330, banana P12,600, and onion P33,240 (FEER 1987:36). Although these figures do not directly indicate profit per hectare, they help explain why landowners have not evicted tenants to grow traditional crops, why mobile capital is interested in fruit rather than grain farming, and why the bulk of the agricultural population is so poor. A brief history of Philippines agrarian systems may help the reader understand the genesis of the present pattern.

Agrarian History in the Philippines

Philippines agrarian history can be divided into phases according to the incorporation of rural land and labor into the world market. Each phase has a characteristic production form, mode of extraction of product, and form of agrarian conflict.

In pre-Hispanic times the archipelago's population was sparse, located mainly on the coasts and rivers. Households produced crops and sea products and domestic manufactures like cloth, pots, and tools for subsistence with some trade between areas in local specialities like salt, iron, gold, and cloth. Households also produced forest and sea products to exchange with Chinese, Japanese, Malay, and Thai traders for luxury ceramics, cloth, bronze gongs, and iron. The people had advanced boats and raided and traded abroad by sea; their ships are reported in Malacca, China, and Java. Coastal communities were small, stateless, and internally stratified. Early Spanish reports describe a lineage chief backed by armed freemen, both ranks served by unfree members of their households and receiving tribute from debt and war slaves with independent households. There are no reports of large farms or of production of crops for long-distance trade.

The Spanish conquest from 1571 superimposed a colonial state and church, both enforcing tribute in product, money, and labor from the population. The chiefs were transformed into colonial headmen, responsible for exacting tribute from their communities, and over the next couple of centuries overt slavery was abolished. Early colonial history is full of reports of families and whole communities fleeing inland from the tribute and forced labor demands of the regime. Spain ruined much of the Asian trade network by destroying native long-distance shipping and restricting access to colonial ports by Asian merchant shipping. For the 200 years from conquest until the end of the eighteenth century Spain did not seek native farm products for export.

Towards the close of the eighteenth century the colony opened to export of tropical agricultural products, setting in train a rapid transformation of the countryside. The galleon trade had declined as Spain's American colonists won their independence and the bullion mines petered out. New shipping routes opened, shipping improved, and the example of the British and Dutch India companies encouraged Spanish colonial governors and colonists and natives alike to experiment with new crops for export. The colonial state set up a tobacco monopoly that between the 1780s and 1880s made large profits through draconian measures against illegal cultivation and trade. Even before the turn of the century Spanish governors began to turn a blind eye to British, American, and other foreign European ships entering the colonial ports flying Asian flags of convenience, in search of cargoes of sugar, rice, indigo, tobacco, and later hemp and copra. Destined at first for one or other leg of the Asian sea trade, by the end of the century export crops were shipped to European and North American markets.

These foreign skippers created effective demand for crops by putting out money or credit through a chain of Chinese-mestizo traders extending from the ports into the hinterland. Accelerating after about 1820 when Manila was opened to foreign ships of any flag, demand for export crops set off migration of peasants from the old seacoast and river towns into the interiors and caused vast areas of forest and grassland to be converted to farm within a couple of generations. Population, which had grown very slowly from 1571 until 1800, began to increase rapidly. The newly rich Chinese-mestizo trading class began to acquire cheap colonial legal title in land by purchase and foreclosure on debt. Augmenting the old principalia of colonial headmen who had accumulated the land of members of their communities unable to meet their tribute, they created a new agrarian system in which landowners who doubled as moneylenders and crop merchants exacted capital gains, rents, debts, and merchants' profit from the product of peasant smallholders and tenants.

In this system the production unit was a small farm under 5 ha, operated by some combination of family, exchange, and hired labor and using technology limited by the peasant farmer's investment capacity. The landowner-moneylender-cum-merchant specified the commodity crop in which he was to be paid rent and debt, advanced food and seed at interest to the farmer, but avoided a wages nexus. He invested in clearing land, irrigation, and land improvement for capital gain only by loaning tools and food to a peasant at interest, recoverable in crop. He rarely invested in improved varieties, machinery, or expended capital in wages, fertilizer, or other annual inputs; he did not supervise the production process, drive the pace of labor, or bear risk. Rather, the rural elite sought to extract capital from agriculture to shift it to education, urban real estate, and politics.

The general strategy of the rural rent capitalist (Fegan 1981), a variety of merchant capitalist, is to reserve capital for trading by avoiding outlaying it at risk in the production process itself in the form of wages and inputs of production. The rent capitalist shifts the costs, management, and risk of production to a dependent petty entrepreneur smallholder-debtor or tenant-debtor, making him bear all labor costs and making the cost of inputs a loan to the operator at interest rather than an expenditure of the capitalist. The rent capitalists' lifestyle involves living in the township to watch price movements, engage in timely buy and sell, and engage in politics and legal matters surrounding debts, mortgages, etc., while pursuing conspicuous consumption. The agrarian system that they dominate generates protest over land-grabbing, conditions of tenancy and level of share rents, rent increases, eviction, the level of interest, the low price at which they value crop delivered against money debt, and abuses of peasant women and the peasant right to vote by the landowner and his staff. The system that took shape in the nineteenth century remains characteristic of Philippines agriculture in the nonperishable low-value crops that occupy the bulk of the agricultural land and population in the archipelago.

By the mid-nineteenth century an agrarian system had developed in which tenant farmers and smallholders living in the villages paid rent and interest to a town-dwelling upper class composed of landowner/moneylender/crop-dealers operating as rent capitalists. By the late nineteenth century these had diversified into the professions, and from this composite upper class were drawn the native political elite of the towns. Limited to the local level despite their wealth and education by Spanish monopolization of provincial and colonial office, it was western-educated members of this dominant class and rural elite whose ideas inspired a lower-class anticolonial revolt against Spain in 1896. Members of this elite seized control of the revolt and then turned to accommodation with the U.S. colonial invader from 1898 when promised access to office under a regime that would encourage agricultural exports. Thus from the turn of the century the native landed class became partners in the U.S. colonial system and its heirs after independence in 1946.

Electoral coalitions of the rural elite, organized on bifactional lines with no ideological or class interest differences (Lande 1965), dominated the independent polity until the 1960s. However, state policy was dominated from the 1950s by the interests of two sets of economic actors. The sugar bloc sought to extend free entry of U.S. manufactures in return for free entry of sugar to the high-priced U.S. market. Import substitution manufacturers were able to manipulate notions of economic nationalism to maintain an overvalued currency, impose duties on manufactures, and help ensure low urban wages by suppressing the rice price through imports.

Agrarian relations varied among crop regions. Sugar was grown under tenancy in the nineteenth century, but a shift to wage labor occurred in Negros in the early twentieth century when large plantations were connected by rail to steam mills and later centrifugal mills (McCoy 1982); in Central Luzon sugar retained a sharecropping system. Copra in the Visayas has been cultivated by smallholders and small share and lease tenants averaging 4.4 ha using some piecework labor; a pure wage-labor system is used only on the largest estates (Tiglao 1983). Rice was produced by a mixture of small owner-operators and tenants, working blocks with an initial working limit of about 4 ha. Censuses from 1903 to 1970 suggest declining operated area, but in the major ricebowls smallholder blocks did not fall much below 2 ha before they were lost by mortgage and landowners resisted tenant blocks being divided into parcels below about 2 ha. Pineapples in Mindanao and in some areas bananas for export, are grown on large estates with the use of wage labor. However, the usual export banana system ties a smallholder to a Philippines corporation by a 'growers' contract' and that corporation to a Japanese or U.S. marketing corporation (David et al. 1983). Abaca (manila hemp) is today predominantly a smallholder and tenant crop, though it was grown on wage-labor estates before the Pacific War.

Successive censuses from 1903 show the proportion of operators who were tenants increasing for all crops, especially for rice, and the type of tenancy shifting from fixed annual rents in kind to sharecropping. By the 1970 census, over 60% of all rice farmers were tenants nationally and over 80% in the Central Luzon ricebowl adjacent to Manila. The predominant form of tenancy from the 1920s was 50/50 sharecropping, with expenses of seed, reaping, and threshing shared by owner and tenant. Landlords lent money to tenants at nominal interest rates of 50% to 100% per crop season for both production and consumption, recovering at the threshing floor debt plus interest in rice at the low harvest price. On large estates landlords from the 1920s collected their share plus debt and interest by compelling the tenant to use estate threshing machines, charging 10% of the gross for the service in the 1930s, falling to 5% by 1970.

In Central Luzon tensions over crop shares, sharing expenses, interest rates, and the use of overseers to police the harvest for absentee landlords precipitated peasant antilandlord unions from the 1920s. Some of these coalesced under Marxist leadership into the KPMP national peasant confederation in 1922. There was widespread tenant-union unrest in the 1930s. The KPMP formed the base for the Hukbalahap (People's Anti-Japanese Army) guerrillas who emerged from the Pacific War with arms, military leaders of some fame, and a Marxist top leadership. From 1946 to the early 1950s the Huk rebelled against the newly independent republic after landowners were allowed to form private armies and began to eliminate peasant leaders and to try to restore powers that owners had lost in wartime. Huk strength was limited to the rice and sugar tenant areas of Luzon, strategically close to Manila. Crushing the rebellion was assisted by a 1953 paper land reform controlling share tenancy conditions. However, landlords in the legislature emasculated this legislation and by refusing to vote funds ensured it could not be implemented (Murray 1972). The specter of peasant rebellion in the strategically located ricebowl continued to concern U.S. advisors and successive U.S. presidents.

There are no useful figures on trends in concentration of landownership, since the agricultural censuses enumerate by farm operators. The increasing proportion of operators who were tenants led many observers to assume that ownership must be becoming more concentrated. However, there are indications that large riceland estates were divided among heirs, parts lost by mortgage or sold to finance higher education and urban investments. Population increase in the context of the end of the land frontier, indivisible farms, and limited urban employment increased the number and percentage of landless households in the villages. We can only infer the figures by comparing numbers of households and farm operators. No figures tell what proportion of nonoperator families were agricultural workers.

From the earliest times for which sketchy information is available, villagers had diverse sources of income. The agrarian regime of a single, risky annual crop and inherited debt plus high-interest loans made it prudent for villagers to raise livestock, engage in home industries, fish, collect forest products or engage in dry-season logging, work at carpentry and construction, transport, trading, as guards, and in lightly populated areas to follow the harvest. By time and region these sidelines vary enormously. What is constant is the remarkable variety of skills and work experiences of the peasantry and the importance of off-farm and nonfarm income sources to households. The rise of a landless class in villages, unable to receive reciprocal labor but needing wages, broke down exchange labor systems that prevail in areas close to the land frontier.

Thus by the 1960s the dominant pattern of agrarian society in the Philippines rice industry was an absentee landlord plus sharecropper system in a crop with low and risky yields, a rebellious tenant class, a growing landless class, and, in rice monocrop areas, stagnant rural towns from which the big landowners had shifted while they continued to drain rents and interest to the city.

Patterns of State Intervention

The state had intervened little in the rice industry; it drew relatively low direct revenues from land taxes, did not invest much in irrigation or research and extension, legislated but did not implement tenancy reform laws, and had tried to intervene in rice marketing to remove Chinese wholesalers from the control of grain trading (Mears 1974). But it was obliged to spend scarce foreign exchange on rice imports. In biennial election years the consumer price of rice was kept down to manipulate public support. State policies favored import substitution manufacturing by overvaluing the peso and imposing duties and import controls. These measures made terms of trade for all agricultural producers poor, and cheap food policies hurt grain producers.

In the 1960s international actors helped to alter state policy. The International Rice Research Institute (IRRI) was set up in the Philippines with initial Ford and Rockefeller Foundation money and later supported by a number of international agencies. The World Bank made credit available for large-scale irrigation works that increased irrigated area by nearly 60% between 1958–1959 and 1967–1968 (Palacpac 1982:59). The U.S. continued to press for land reform as its concerns about rural insurgency deepened with the Vietnam War, and after withdrawal the strategic importance of U.S. bases in the Philippines was increased by Russian access to the Cam Ranh Bay complex. International concern about world food supplies had wide ramifications in the availability of credit for increasing rice production.

Internally, economic nationalism peaked in the late 1960s. Emphasis on import substitution manufacturing deepened concern about rice self-sufficency and helped create a climate of opinion in Manila favorable to land reform. At the same time, a cadre of U.S.-trained technocrats was available to manage economic policies.

After his unprecedented reelection in 1969, President Marcos was persuaded that the recent advances at IRRI in rice varieties and technology could solve a number of national problems. It had become widely believed in academic and government circles that sharecropping contracts left the tenant with no incentive to increase harvest since any gain would go to the landlord in crop shares and payment of old debt. Thus share tenancy seemed a barrier to increases in rice productivity, as well as a cause of insurgency. Meantime, the major legal political opposition was drawn from very powerful families some of whose power bases lay in large rice and corn estates. There were peasant marches from Central Luzon demanding land reform, supported by a daily picket around the Congress of students, priests and nuns, unionists, and other urban elements supporting the peasants.

In September 1971 Marcos approved an amendment to the land reform (restricted to rice and corn land). This reform made sharecropping 'illegal' and exhorted a nationwide system of fixed rents in kind set at 25% of the average harvest net of seed reaping and threshing, but it made the tenant responsible for all costs. To allow tenants cheap credit to use the new cash inputs he set up institutions charging 12 to 14% interest, below the market rate of about 30% and well below the informal rural credit rate of 10% per month. Credit funds were, however, limited. Landowner resistance to leasehold was massive and largely successful in maintaining share tenancy or forcing rents higher than the legal rate.

In October 1972, a month after declaring martial law and with no Congress to impede legislation, budget, or implementation, Marcos went a step further. Presidential Decree No. 27 instituted a 'land to the tiller' program that made leasehold compulsory in rice and corn land nationwide. It also provided for tenants of landowners with over 7 ha to buy their land over fifteen years by paying 2.5 times the annual normal net harvest plus 6% interest, a figure designed to make amortization installments about equal to leasehold rent of 25% of the net harvest. The first OPEC oil price rise of 1973 and an international shortage and price rise for nitrogen fertilizer and rice in 1973–1974 precipitated a crash program for small-farmer credit. This was channeled through the Masagana-99 scheme, which made it compulsory to use modern varieties and fertilizer.

The land reform posed a major threat to the big landlords. Marcos seems to have calculated that a large proportion of riceland was owned by a few very big owners, so that the reform would make many friends and few enemies (and those already his opponents). Landowner resistance included evading the law by switching crops, evicting tenants, dividing title, and using a number of measures to drive up the valuation or rent of the land as a minimum while trying to obstruct, delay, and subvert the reform. Meantime peasant organizations that might have kept up pressure for speedy and thorough implementation were broken up, and media that might have revealed corruption, obstruction, and evasion were suppressed (Wurfel 1977).

Marcos wavered when it turned out that a greater proportion of land was owned by small and medium owners than expected and that these included the professionals, officials, officers, and 'economic middle class' in general on whom the regime depended for implementation of its overall policies and who Marcos hoped might become a new constituency for his regime after destruction of the rural oligarchs. He allowed the reform to be watered down and did not punish resisting landlords.

Thus the reform gave land-purchasing rights to fewer tenants than expected, but it did give to all who survived the evictions secure tenure and fixed rents, just at the time when the new technology was about to take off. Although landlords tried to avoid, obstruct, and delay the reform and to push up land valuation and land rent above the legal formula, the regime did protect tenants against evictions after the first weeks, and rents had become fixed by about 1976. As it happened, this was just when the Green Revolution technology made the greatest gains in yield, so that some owners did not participate in the windfall.

The effects of the reform on agrarian differentiation were real but limited. The reform was conservative in that it was concerned solely with relations between the existing landowner and tenant and did not confiscate owners' land without compensation or threaten the tenants with redistribution of land among the whole village population in order to benefit the landless. It circumscribed private property rights by directing that large estates be sold to their tenants at a price below market value, while at the same time recognizing them in directing that landlords were to be paid.

As tenants came to recognize that security of tenure was real, many ceased to pay amortization and lease rent or paid less than the due amount while also ignoring old debt arrears to landowners. The state has not allowed eviction for nonpayment. Although the reform set a target farm size of 5 ha for rain-fed and 3 ha for irrigated land, the reform agency quickly realized that a farm equalization program would be politically dangerous. It would alienate the big peasants on whom it depended for village control since village political leaders were typically drawn from farmers with oversize blocks. In effect Marcos retreated to a reform-in-place, preserving existing inequalities between farmers. It should be emphasized that the reform was restricted to tenanted rice and maize land so that the benefits of tenants there did not impede the processes by which smallholders continued to lose land to landgrabbers and moneylenders in other crops or increases in rents there. Wurfel (1977) argues that more nongrain smallholders lost land in the 1970s than the grain farmers who gained purchase rights from the reform. Finally, the reform was weakly implemented, if at all, outside of the Central Luzon, Ioilo, and Isabela ricebowls; it had little impact in the Eastern Visayas and Mindanao.

Cheap institutional credit released by the government encouraged farmers in favored areas to adopt the new technology quickly. By 1980 over 85% of irrigated area and 70% of rain-fed land were under modern varieties, and fertilizer use rose from 22 kg/ha in 1960, 73 kg/ha in 1970, to 147 kg/ha in 1979 (Palacpac 1982). Farmers proved bad debtors, partly because rice is a risky crop in the Philippines where the ricebowls are ill-drained floodplains on which typhoons in the wet season cause great damage while the bulk of the land remains rain-fed and subject to drought.

The Masagana-99 credit program at its peak in 1974–1975 covered 40% of the area, although by 1977–1978 this had fallen to 10% (David 1979). Those who continued to obtain credit were the minority of better-off farmers in irrigated areas. However, a number of rural banks were set up to channel the credit. Many were owned by large landowners. The rediscounting and loan guarantee system run by the Central Bank created conditions where some bankers found bad debts of farmers profitable. The regime's political imperatives of increasing rice production and pacifying the peasantry ensured that political rather than bankers' criteria determined the volume of loans and who repaid them. Farmers were in effect trained to be bad debtors, many accumulating debts at several outlets. They quickly realized that the poor collection mechanisms and lack of sanctions made it easy to take the credit and run.

A divergent attempt to ensure rice production and procurement was the Corporate Farming Program. Government Order No. 47 of 1974 directed corporations with over 500 employees to grow rice at the rate of 1 ha per seven employees. It provided two options: opening land at the frontier in Mindanao or working through contracts with existing small farmers, under which corporations would supply farmers with inputs, management, and supervision (Tadem 1978). In the event, even corporations experienced in other crops abandoned rice growing when state pressure eased, as the rice and fertilizer crisis of 1973–1974 receded.

The government did not subsidize fertilizer to the farmers except during the crisis years 1974–1975. Domestic prices from 1971 to 1983 ran about 10% above world prices at the border. Huge subsidies from farmer and government went to inefficient domestic manufacturers of fertilizer (David and Balisacan 1981). The fertilizer/rice-price ratio in the Philippines is among the highest in Asia, contributing to the comparatively low yields in an all-Asia context (Table 6.1).

In 1972 the government reorganized the Rice and Corn Administra-

TABLE 6.1. *Fertilizer/Rice-Price Ratios: 1981*

Country	kg N/kg Paddy
Bangladesh	1.76
Burma	1.80
India (Coimabatore)	2.71
Indonesia (Central Java)	1.62
Japan (1979)	0.45
Korea (South)	1.19
Malaysia	1.77
Pakistan (1979)	3.61
Philippines	3.68
Sri Lanka (1980)	1.08
Taiwan	1.33
Thailand	3.35

Source: Coxhead (1984).

tion as the National Grains Authority (in 1981 renamed the National Foods Authority—NFA) to intervene in the grain trade. It had conflicting objectives—ensuring supplies and maintaining floor prices as an incentive to farmers and ceiling prices to favor consumers. The NFA used its monopoly of international grain trade to import rice in anticipation of deficits and restrict exports, in effect keeping the market flooded. Domestic prices followed the trend of world prices, slightly above them in the 1960s but below them in the late 1970s for comparable quality (Unnevehr 1982). But from the 1970s until the 1983 devaluation the system of manufacturing protection resulted in a 20% to 30% overvaluation of the peso. Domestic prices would have been above world prices had this policy not applied (David 1983). The NFA maintained the ceiling price but not the floor price against seasonal fluctuations, so that all rice producers were penalized, particularly small farmers obliged to sell quickly at the low harvest prices.

Increases in rice production ran at 8% per year through the 1970s in response to the combination of new varieties, fertilizer, and irrigation investment. But state intervention in markets, together with increased harvests, caused a steady decline in the real price of rice and in the terms of trade for rice producers (Table 6.2). From a base of 1 in 1972, rice producers' terms of trade fell to 0.59 in 1980. Meantime the prices of

TABLE 6.2. *Philippines Terms of Trade for Rice Producers*

Year	Rice price index	Nonfood price index	Terms of trade
1972	100	100	1
1973	114	126	0.90
1974	171	166	1.03
1975	181	182	0.99
1976	173	199	0.87
1977	178	221	0.81
1978	170	240	0.71
1979	186	281	0.66
1980	199	340	0.59

Source: Coxhead (1984).

agricultural inputs rose slower than the consumer price index (CPI) but nevertheless faster than the price of paddy rice. This squeezed farmers between low product prices and high input prices. The effect was exacerbated as the nation increased its international indebtedness so that cheap institutional credit dried up in the late 1970s, forcing farmers to the informal credit market where rates run at 10% per month.

The overall effects of relative price changes were to make farming unprofitable. By 1983 areas of rain-fed and flood-prone land were abandoned as farmers failed financially. Farmers accumulated large arrears of amortization and rent to landowners, owed irrigation fees, defaulted on debts for machines and inputs, and were obliged to reduce fertilizer. Yields and the national harvest fell. In 1984 and 1985 the Philippines resumed importing rice. In Central Luzon a new farmers' organization in early 1984 was mounting demonstrations to protest the fertilizer and paddy prices set by the state, instead of the conditions of tenure. They threatened to revert to traditional varieties that need low cash inputs and advised farmers to grow only enough for home consumption (Bartolome 1984). Though the scale of the organization is unclear, its advice coincided with what some farmers were doing anyway in response to the price squeeze.

Herdt and Gonzalez (1980) indicate that economies of scale would not insulate the 500-ha corporate farm from these unfavorable price changes. Using the same level of physical inputs and outputs found in field studies in 1974, they calculate that relative price changes by 1980 would turn the large farm's 1974 profit of P608 per hectare to a loss of P636. Although differences in irrigation, farm size, mechanization, and input levels gave different estimated profitability, their calculations show a decline in net farm income under most regimes over that period.

Aside from legal barriers posed by the land reform and the problems of peasant resistance and labor organization they have encountered, there was little incentive for agricultural entrepreneurs to enter or remain in rice farming, for landowners to displace the increasing number of tenants who defaulted on rents, or for richer villagers to take over the farms of their neighbors via debt.

If the relative prices protected small rice farmers from losing land by removing capitalist farming competition, they also forced cost cutting. As it happened the opportunities were in the area of rent and labor costs. Rural wages in the 1970s rose slower than the CPI but faster than the price of paddy or the price of labor-displacing machinery and chemicals. Coxhead (1984) calculates that the price of nitrogen fertilizer, deflated for the price of paddy, was stable throughout the period 1975 to 1983. Similar relative price changes occurred for herbicides, but the rural wage rate, deflated for the paddy price, doubled in the same period. In the mid 1970s IRRI developed and promoted a system of broadcasting germinated seed direct onto the paddyfield mud and controlling weeds with herbicides. In favorable locations farmers switched in the late 1970s to the new system in order to eliminate the labor cost of uprooting seedlings, hand transplanting, and weeding plus the problems of labor organization and discipline.

The IRRI-designed two-wheel hand tractor had largely displaced buffalo from land preparation in Central Luzon and Laguna from the early 1970s after several IBRD and Central Bank credit programs for farm mechanization, plus the readiness of private banks to accept the land reform Certificate of Land Transfer (CLT) as security for machine loans. The labor displaced was mostly that of the farm family, for few landless owned buffalo. But the use of herbicides displaced hired labor, particularly that of women whose main tasks in the rice cycle were transplanting and weeding. This made landless families more dependent on earnings from reaping and threshing.

Farmers refused to use the cumbersome estate threshing machines from about 1978 in Central Luzon after the 1972 land reform broke the power of landowners to compel them to bundle, haul and stack the reaped paddy and then wait several weeks for the machine to be available. The political demise of the big machines gave laborers some bargaining power for a year or so, particularly where harvests were bunched over a wide area because of the irrigation schedule or bad weather. However, IRRI's highly mobile axial-flow threshing machine design had been released and banks were ready to accept CLTs as security for its purchase. Rural entrepreneurs rapidly spread the new machine, which allowed farmers to eliminate making and hauling sheaves, to thresh each paddy field separately, and get immediate access to the grain. Labor's share dropped from 16.6 or 20% of the crop to 11% immediately, lower in some areas, as its harvest tasks were cut back to reaping and feeding the machine. By 1981 IRRI's light reaping machine was spreading in provinces near Manila, where it reduced the need for hired labor in reaping. With that will go hired labor's last major task in the rice production cycle, their bargaining power in remaining tasks, and their major source of income (Fegan 1981). Since 1982 the design has been in the public domain, although some technical problems remain to be solved.

The prospects for landless laborers seem bleak: they will be largely displaced from the rice production cycle by machines and chemicals. IRRI innovations, in the context of government policies that make capital artificially cheap, have marginalized landless agricultural workers. Its policy has been to design for the private benefit of the farmer and owner of capital, whatever the social consequences and despite the high social cost of capital and low cost of labor or the national cost to a debtor nation importing petroleum fuel and chemicals. The prospect is that displaced rural labor will seek alternative work in the city, exacerbating Manila's slum problem and depressing urban wages.

Both of the essays that follow rely on restudies of villages in Central Luzon, each located within 90 km of Manila. Central Luzon accounted for 20% of the national area under rice and about 27% of national production in 1969–1971 (ILO 1974:tables 97 and 98). It is the best-studied area in the Philippines because of its proximity to Manila and to the IRRI, rapid spread of agricultural innovation, and history of agrarian unrest.

However, Central Luzon is hardly 'typical' of rice-growing areas in the Philippines. It has the largest proportion of irrigated area, the highest rate of tenancy, the highest yields, and, because of the nearby labor markets of Manila and U.S. bases, the highest real rural wages. Agricultural risk may be higher than in better-drained areas or places south of the usual track of typhoons. Central Luzon has better roads, higher farm prices for products because of proximity to Manila, better agricultural extension and access to institutional credit, and was a target area for the 1972 land reform. Its population has higher rates of literacy and better education than average, is more politically conscious, and is served by more radio, television, and print media than other regions, while the language of about half its area is Tagalog, the basis of the national language.

Had studies been available from the Eastern Visayas (Samar-Leyete), they would have shown a much poorer and less technically advanced rice agriculture oriented to subsistence production. Studies from Iloilo would have shown a recent expansion of modernized production in a predominantly rain-fed area. The ricebowls of Davao and Cotabato in Mindanao are areas of postwar in-migration by Christian settlers, racked by conflict between Christian and Muslim and of communist NPA versus the armed forces. Farm sizes are larger and tenancy rates lower than in Central Luzon, but trending towards the national average, while paddy prices are lower, inputs prices higher, technology more traditional, and government services including extension and credit poor. Many areas mix rice and coconut cultivation. Poor market, extension, irrigation, and general services are also characteristic of the postwar frontier provinces of the Cagayan valley of Luzon, where NPA/army clashes are common, while land and labor are less devoted to rice as a monocrop.

Conclusion

Since classical times no state has been prepared to leave to market forces its political imperatives of extracting grain from the producers by means that avoid rural rebellion while ensuring food supplies to the populace at prices that do not provoke urban riots. States today intervene in agricultural markets, using diverse policy instruments to promote industrialization and grain self-sufficiency and to win political support. The nation-state has been joined by international scientific organizations like IRRI, by international lending agencies like the IBRD/WB and ADB that promote irrigation and agricultural modernization policies, and by IMF and consortia of creditors that influence national macro-economic policies. Markets are manipulated by multinational cartels like OPEC that affect world fuel and agrochemical prices and by nation-states and organizations like the United States, Japan, Malaysia, and the EEC that subsidize their grain producers and then dump surplus or withdraw from purchasing, thereby depressing commodity prices. The relative costs of technologies that compete with labor and animal power are affected by worldwide technological changes, commodity prices, and availability of credit.

In the Philippines the political economy of the rice industry has meant the crop has not been profitable for most producers since the early 1970s, least of all for large farms using hired labor and large inputs of fuel and chemicals. Although the land reform of the Marcos state broke the control of large landowners over their tenants, it is probable that had the crop been profitable then landowners would have devised means to recover their land, that corporate farming would have persisted and spread, and that richer peasants and other rural capitalists would have been able to gain working control of substantial areas of land. That they have not, though the means were there in the indebtedness of farmers and venality of officials, confirms that there were more profitable investments. It is ironic that the very forces that kept the small farmers poor also preserved their farms. It is tragic that related forces did not preserve the jobs, income, and dignity of the landless laborers. Although the collapse of OPEC, leading to falling world fuel and agrochemical prices, will reduce the hard currency price of nitrogen fertilizer and chemicals, devaluation of the peso and the elimination of subsidies at the behest of the IMF and a bankrupt treasury have kept the farm cost of chemicals and machines high. With a continuing low price for paddy, capitalist farming cannot be expected to compete for land with the smallholding farmer. It is more likely that the post-Marcos regime's conflicting imperatives to supply the populace with cheap grain and maintain rural peace will constrain it to keep rice farmers' profits low, subsidize fertilizer credit, and at the same time attempt to complete the land reform. The prospects, then, are that the smallholding will be preserved as the unit of production for rice.

References

Bartolome, N. (1984). "Serious Threat from CL Farmers." *Malaya*, October 4.

Coxhead, I. (1984). "The Economics of Wet-Seeding: Inducements to and Consequences of Some Recent Changes in Philippine Rice Cultivation." Unpublished dissertation, Master of Agricultural Development Economics, Australian National University, Canberra.

David, C. (1979). "Structure and Performance of Rural Financial Markets in the Philippines." Occasional Paper no. 589, Agricultural Finance Program, Ohio State University.

——— (1983). *Economic Policies and Philippine Agriculture*. Philippine Institute for Development Studies, Working Paper 83-02.

David, C., and A. Balisacan (1981). "An Analysis of Fertilizer Policies in the Philippines." Mimeo. Workshop on the Redirection of Fertilizer Research, Tropical Palace Hotel, Manila.

David, Randolf, et al. (1983). "Transnational Corporations in the Philippine Banana Export Industry." In Third World Studies Program, *Political Economy of Philippine Commodities*. Quezon City: University of the Philippines.

Far Eastern Economic Review (FEER) (1987). "Shifting to Cash Crops for Higher Income." *Far Eastern Economic Review*, March 5.

Fegan, B. (1981). "The Grim Reaper: A Decade of Harvest Increasing and Income Redistributing Changes in Central Luzon." Paper presented at IRRI, Los Banos.

Herdt, R. W., and L. A. Gonzalez (1980). "The Impact of Rapidly Changing Prices on Rice Policy Objectives in the Philippines, 1980." Mimeo. IRRI Agricultural Economics Department, Paper no. 80-16.

ILO (1974). *Sharing in Development: A Program for the Philippines*. Manila: International Labor Office.

Lande, C. (1965). *Leaders, Factions and Parties: The Structure of Philippine Politics*. New Haven: Yale University Southeast Asian Studies.

McCoy, A. W. (1982). "A Queen Dies Slowly: The Rise and Decline of Iloilo City." In Alfred W. McCoy and Ed. C. de Jesus, eds., *Philippine Social History*. Quezon City: Ateneo de Manila University Press.

Mears, Leon, et al. (1974). *Rice Economy of the Philippines*, Quezon City: University of the Philippines Press.

Murray, F. J. (1972). "Land Reform in the Philippines: An Overview." *Philippine Sociological Review* 20: 1–2.

Palacpac, Adelita (1982). *World Rice Statistics*. Los Banos, Laguna: Department of Agricultural Economics, IRRI.

Stickney, R. E., et al. (1984). "Introduction of the CAAMS–IRRI Mechanical Reaper in the Philippines." Manila: Agricultural Engineering Department, IRRI.

Tadem, Eduardo (1978). "Peasant Land Rights and the Philippine Corporate Farming Program." *Philippine Social Sciences and Humanities Review* 42:1–4.

Tiglao, Rigoberto (1983). "The Political Economy of the Philippine Coconut Industry." In Third World Studies Center, *Political Economy of Philippine Commodities*. Quezon City: University of the Philippines.

Umehara, H. (1983). Green Revolution for Whom? In A. J. Ledesma et al., eds., *Second View from the Paddy*. Quezon City: Ateneo de Manila University Press.

Unnevehr, L. J. (1982). "The Impact of Philippine Government Intervention in Rice Markets." Mimeo. Paper presented at the IFPRI Rice Policy Workshop, Jakarta, 17–20 August.

Wurfel, David (1977). *Philippine Agrarian Policy Today*: Implementation and Policy Impact. Occasional Paper no. 46. Singapore: Institute of Southeast Asian Studies.

Chapter Seven

The Saudi Connection: Agrarian Change in a Pampangan Village, 1977–1984

CYNTHIA BANZON-BAUTISTA

This study tries to identify some of the mechanisms propelling agrarian households in different social and economic directions in the rice village community of Santa Lucia in Pampanga, Central Luzon. In particular we will examine the role of access to opportunities within and outside rice agriculture in influencing the pattern of agrarian differentiation during the years 1977–1984, a period of general economic decline in this region as in the Philippines.

We will argue that these patterns cannot be accounted for solely or primarily by reference to the technical and social changes related to agrarian reform and that their explanation must take into account accumulation resulting from wage-labor opportunities arising outside the village, namely in the Middle East. Since Sta. Lucia is one of many labor-exporting agrarian villages in Luzon and other regions where income from agriculture is no longer the primary source of income for many or even a majority of households, the arguments developed here may be of quite wide significance in shedding light on patterns of differentiation in other Philippine villages.

The study is based on employment and production data obtained from the same sample of households at two points in time, 1977 and 1984. As discussed more fully later, restudies of this type for the purpose of understanding mechanisms of agrarian change have both advantages and limitations compared to other approaches.

Santa Lucia: Agrarian History

Santa Lucia, the village under consideration, is located in the town of Sta. Ana, Pampanga, about 80 kilometers from Manila. It is thirty minutes by bus from Mt. Arayat, the famous hiding place of peasant rebels during the Huk rebellion of the 1950s and about the same distance from the municipality of San Luis, the former residence and stronghold of Luis Taroc, the leader of the rebellion. To the west is the municipal center of Sta. Ana and to the east are the swampy lands of Candaba. Although some farms in low-lying areas close to Candaba are sometimes flooded, most Sta. Lucia farms are protected by a flood control dike built before World War II and reinforced after 1972. This partly explains the migration of some Candaba residents into the area.

The village has been settled since at least the latter part of the nineteenth century and possibly much earlier, since it is in the immediate vicinity of the municipality, Sta. Ana, whose origin has been traced by Larkin (1972) to the expansion of older settlements from Pampanga to outlying areas in the 1750s. Hacienda formation in the village occurred in the early 1900s when a municipal judge acquired thirteen land titles amounting to about 302 hectares. The hacienda became a dominant feature of the village since the two other landholdings made up only 13 hectares. From the late 1930s up to the late 1960s, it was characterized by absentee landlordism, a system of overseers, and share tenancy.

By the mid-1960s, the organization of the hacienda had begun to disintegrate. The death of the landowner and the inability of his widow to manage the estate by remote control further eroded the relations between the new owner and the tenants. Under the leadership of the former overseer, who was himself a tenant, the farmers struggled to change their tenurial status from share tenant to lessee. By 1968, leaseholding rights were granted to share tenants under the terms of the 1963 land reform code. With the implementation of the 1972 land reform program, about 50% of the leaseholders received certificates of land transfer which entitle them to ownership of the land upon full payment of amortizations. Most farmers, however, have not begun amortization payments and rental collections have not been seriously enforced, because no land valuation agreements have been reached with the landlord.

Most of the lessees in the sample paid only token rents, if at all. In 1977, all farm operators in the village had the status of tenants; all the landowners were absentee, and none had turned to direct operation of large farms using machines and hired labor.

Many of the older residents of Sta. Lucia have had a history of involvement in organized and militant peasant struggles. In the late 1930s, a number joined the peasant strikes and uprisings which characterized this region. (Pampanga had one of the highest frequencies of peasant strikes in the 1930s and 1940s; cf. Kerkvliet 1977.) They also took part in the postwar peasant movement. A few claim to have undergone political education in the 'schools' of Mt. Arayat at the height of the Huk rebellion.

Organized struggles, albeit in different forms, continue to be manifested in the village. At least three incidents reflect the effects of such struggles. In 1968 when the hacienda owner withdrew his mechanical pumping unit which had drawn water from a reservoir to retaliate against peasant demands to end share tenancy, the village residents converted the reservoir into a communally owned fishpond. The funding and development of the fishpond were facilitated by an organizer from a private development agency. Later, in their struggle to retain as much rice as possible, farmers in Sta. Lucia initiated a collective project with farmers from other villages and installed a rice mill using external funds. The mill run by a farmers' federation has enabled peasants to increase their recovery rates of milled rice and also to retain the by-products of milling. Finally, in 1977, landless farm laborers occupied and subdivided 40 hectares of undistributed idle land belonging to the hacienda among themselves and transformed it into riceland. Motivated by poverty and probably by the prospect of gaining land through the state's land-to-the-tiller policy, a number of landless farm laborers began clearing and preparing a portion of the idle land. While the landless were not formally organized, the village council and the leader of a farmers' association intervened to prevent a mad rush for the remaining idle land. Without consulting the absentee landlord, they drew up a list of forty individuals and conducted a lottery to determine the assignment of land parcels. Among the forty landless on the list was the owner of a large store and a local labor contractor who wielded influence in the municipality. The new farmers later signed an agreement formulated by the landlord, undertaking to return the land if she should ask for it at any time. The farmers paid rent for the first four years but stopped paying when the owner refused to issue receipts. The landowner's refusal may have been an attempt to avoid future land transfer. The issuance of receipts is an explicit recognition of the occupancy rights of the forty landless. In light of the state's land reform program, this recognition may have been perceived as a basis for subsequent land transfer operations. It is interesting to note that in 1983, the landowner attempted to borrow from the rural bank using this land as collateral but was refused a loan.

The transformation of this idle land into forty small rice farms caused an abrupt change in the pattern of landholdings, increasing the number of farms by about one-half between 1977 and 1978 and at the same time greatly increasing the prevalence of small farms; farmholdings of less than 2 hectares were about 25% of the 82 farms in Sta. Lucia in 1977, but about 50% of the 126 farms in 1978. Conversely, the number of larger farms of 4 hectares and above—which has decreased absolutely from 17 to 12 due to the sale or leasing of land in the covert land market, for reasons discussed below—has declined proportionately from 21% to 10% of the total due to the addition of the large number of small-farm households. Shifts in landholding patterns, therefore, reflect a combination of economic and political changes.

Although sugarcane is said to have been cultivated in earlier times, rice is now the dominant crop in the village. High-yielding varieties from the International Rice Research Institute (e.g., IR-26 and IR-30 in 1977 and IR-36 and IR-42 in 1984) have gained much popularity among farmers since their introduction in 1970. However, the yields as reflected in the wet-season harvests of sample farmers in 1977 and 1984 have not been high. The average yield per hectare during the 1978 wet-season harvest was only 60 cavans (approximately 3 metric tons) per hectare; in 1984 yields had dropped slightly to 58 cavans, due to problems of tungro infestation. Because of natural disasters (e.g., tungro and rat infestation, flooding in low areas) and the increasing costs of chemicals and fertilizers in the 1970s, a number of farmers have now begun to combine modern with more traditional rice varieties which do not require chemical inputs; about 12% of the sample farmers used traditional with modern varieties in 1984.

Until 1968, water supply depended on rain supplemented in part of the area by water drawn from a reservoir by a mechanical pumping unit owned by the landlord. After the landowner withdrew his machine in 1968, peasants began to fend for themselves. From 1972 onwards, they either borrowed money from the rural banks or used part of their own earnings to acquire individual irrigation pumps. Some formed small irrigation associations which collectively acquired pumping units. Compared to areas like Laguna with large-scale gravity irrigation systems, water supply and costs are still a problem for some farmers. Because the pond dries easily, only a low proportion of farmers plant dry-season crops.

The adoption of the new rice technology at first increased the volume of work for farmholding households, leading to greater demand for hired laborers. To save on costs, farmers hired workers only for transplanting, pulling of seedlings, harvesting, and postharvest tasks (e.g., gathering and stacking the harvest for machine threshing). However, based on sample data, some changes in techniques have affected the farmer's demand for hired laborers. About 40% of the sample farmers in 1984 broadcast pregerminated seeds into the mud, eliminating the need to hire laborers for uprooting seedlings, transplanting, and weeding. Moreover, at least three farmers in the sample have recruited permanent laborers from Bicol to replace family labor as well as some hired laborers.

Farm employment opportunities for landless workers are further constrained by the prevalent *atorga* harvesting arrangement between farmers and farm laborers. This arrangement (similar to the *gama* arrangement now widespread in Laguna but in this case involving gangs of laborers rather than individual households) specifies that the laborers perform the pulling of seedlings for free in exchange for the right to cut, gather, and stack the unhusked rice for machine threshing. The group then receives a sixth of the harvest of which 5% goes to the threshing machine operator. The *atorga* contract (which excludes from the harvest those who are not members of the *atorga* gangs) emerged in a nearby municipality in 1970 with the use of the new rice technology. There the seedlings, which were shorter, harder to uproot, and required less time in seedbeds, increased the demand for hired labor in seed-pulling. Migrant workers from another village ensured their access to employment in the most remunerative harvesting phase of production by offering to pull seedlings for free in exchange for harvesting rights. Since this arrangement reduced costs for farmers, it was institutionalized in the municipality. Santa Lucia adopted the arrangement only three years later (because of pest infestation in 1971 and a typhoon in 1972). Since then, there have been only a few instances when nonmembers of the exclusive harvesting groups were allowed to join in the harvest; in general, harvesting opportunities for other landless laborers exist only on farms

TABLE 7.1. *Primary Occupations of Household Heads: 1977 and 1984, Sta. Lucia*

Occupation	1977 N	1977 %	1984 N	1984 %
Rice farmer	82	26	114	25
Farm labor	59	18	46	10
Vegetable grower	8	3	2	0
Construction				
Carpenter, mason, tinsmith	47	15	108	23
Heavy equipment operator	10	3	10	2
Electrician, welder	11	3	9	2
Foreman	8	3	1	0
Contractor	7	2	2	0
Overseas worker/employee	17	5	78	17
Services				
Driver, conductor, tricycle rig driver	24	7	50	11
Mechanic	2	1	4	0
Merchandise seller, businessman	12	4	16	3
Employee, policeman	10	3	6	1
Tailor, photographer, barber, printer	7	2	2	0
Bakery or textile worker	5	2	2	0
Piggery or poultry worker	4	1	8	2
Thresher contractor	1	—	2	0
Teacher, professional	5	2	5	1
Total employed heads	319	100	465	100
Unemployed	6		2	
Old, widowed, sick	19		11	
Total heads	344		488	

which retain traditional open harvesting arrangements. It should be noted that while *atorga* is still prevalent in the sample, the change to direct seeding by some farmers has reduced the number of farms under *atorga* contracts.

Although Sta. Lucia may still be considered an agrarian village community, agriculture is no longer the primary income source for most household heads, as may be seen in Table 7.1. In 1977, half of the total employed heads reported their primary occupation either in construction work or in the service sector, and by 1984 this proportion had risen to almost two-thirds. Meanwhile farm labor as a primary occupation of household heads has declined. Many of these heads were absorbed in the construction sector as carpenters, masons, and tinsmiths; a few joined the ranks of overseas workers whose numbers increased from 1977 to 1984, as we shall see below.

Because of its significant role in explaining many of the changes in the economic and social conditions of the sample, a word must be said about the phenomenon of overseas migrant labor. Since the second half of the 1970s, export of manpower has become one of the key strategies undertaken by the state to earn badly needed foreign exchange and to provide employment for the underemployed population. This period coincided with the construction boom in the Arabian Gulf regions. Because Pampanga is known for its carpentry work, labor contractors recruited many workers from the area; in 1979, Pampanga had the largest number of Filipino contract workers abroad (Arcanes 1984). In one Pampanga village, almost all families have members working overseas.

Changes in the Sample Farm Households: 1977–1984

The processes of agrarian change in Sta. Lucia can be partially inferred from changes in the social and economic conditions of the same farm households through time. This section describes the changes in productive assets, and in the levels and sources of household income between 1977 and 1984, and explores some of the causal mechanisms which brought these changes about.

In 1977, heads of sixty-three farmer and thirty farm labor households (77% of all farmer households in the village and 51% of all farm labor households in the village in that year) were interviewed. The same households were revisited in 1984. In households whose heads had died, were too old, or had migrated temporarily for overseas contract work, the wives or heirs were interviewed; information was also obtained about the farms which were completely lost by the respondents during the eight-year period. Key informants helped to verify the data for any households which showed dramatic changes during the seven-year interval.

Restudies of the same sample of households for the purpose of understanding mechanisms of agrarian change have both limitations and advantages compared to other possible approaches. One problem is that the respondents covered in 1977 no longer constituted a representative sample of farmer and landless farm labor households when reinterviewed in 1984. A further limitation in this study is that the occupational data are limited to primary occupations of household heads, which means that we do not have a complete picture of the patterns and combinations of income-earning activity at the household level for either 1977 or 1984. However, the strength of the method lies in the possibility of tracing the fortunes of specific households and in particular of observing the mechanisms by which accumulation or loss of land and other assets has occurred.

Of the 63 heads of farm households in the original sample only 50 were still engaged in farm production in 1984: 2 had become farm laborers, 1 was a self-employed businessman in the municipality, 4 were overseas workers, 1 was a construction worker in Manila, 2 old men had bequeathed their farms to other family members, and 3 had died. The 1977 data for old or deceased respondents were compared with the 1984 data for the household head who maintained the respondent's house and a part of his farm.

Of the 30 landless farmworker household heads in the 1977 sample, 9 had become farmers in 1984, being among the 40 landless residents who received 1-hectare rice farms in 1978 in the manner described above. Moreover, 16 continued to work as hired laborers with nonfarm sources of income, 2 were working overseas, 1 was a carpenter in Manila, and 2 had brought their families to the national capital in search of better employment opportunities. These two households are excluded from the 93-household sample in the discussion which follows.

Changes in Productive Assets

As noted earlier, tenant smallholdings have dominated the barrio since the formation of the hacienda at the beginning of this century. Share-croppers then were assigned land parcels ranging from slightly less than 1 to 5 hectares, with about 40% cultivating between 1 to 3 hectares. Land reform merely froze the system of smallholdings. Although the ricelands assigned to farmers remain small, there have been interesting changes in the sizes of the landholdings of the farms in the sample.

The acquisition by nine landless households of 1-hectare ricelands each is one of the more dramatic changes in the seven-year period. The landless who took part in the informal takeover of the hacienda's idle land may have done so as a coping strategy in the face of declining employment opportunities and the dwindling natural sources of subsistence (e.g., field frogs, fish). It is possible also that some of these landless harbored hopes that the state, with its public adherence to the 'land to the tiller' philosophy, would eventually grant them tenure as permanent lessees. This may account for their initial eagerness to pay the annual rents deemed necessary for establishing their claims to the land, in contrast to the tenured farmers in the village who have been remiss in their land rental payments. It may also account for the landlord's refusal to issue receipts to those who paid. Whether the implicit arrangement between the new farmers and the landlord will continue remains to be seen. However, given the level of organization of the former, the tacit support of village officers, and their political connection with municipal officials, one of whom also acquired a 1-hectare parcel, it may be difficult for the landlord to regain control over this land.

Besides the emergence of small farmers from this segment of the landless, other important changes in farm sizes occurred. Within the seven-year period, about one out of three farmer households experienced significant reductions in landholdings. Of these fifteen households, five bequeathed part of their farm to a son or grandson; this includes those whose lands were subdivided as a result of the death or physical incapacity of the head. The remaining ten sold or leased land in the covert land market, but only in one case were severe economic difficulties the cause. The rest leased or sold part (and in four cases all) of their land in order to raise the necessary capital for a business venture or to cover the costs of obtaining overseas employment for themselves or their families. 'Loss of land' in these cases, therefore, seems to have been part of a strategy of diversification of income sources, not always provoked by distress.

Two farmers who lost all of their 3-hectare landholdings represent a study in contrast. One was able to accumulate enough capital from a retail store, from rentals of a small machine thresher, and from the sale of his 3-hectare farm to finance a funeral parlor in the municipality. The other, hoping to land a 'Saudi' job (as employment in any country of the Arabian Gulf region is called), became a victim of illegal labor recruiters and lost his farm without the prospect of a better source of income. Instead of moving up in the village hierarchy, he joined one of the *atorga* harvesting groups, occasionally engaging in construction work during the slack season. While the household of the first farmer is accumulating, that of the other is struggling on the margin of subsistence.

Of all lands lost from 1977 to 1984, about half were leased or sold in the process of obtaining Middle East work contracts for the head or a member of his family. Except for the single victim of illegal recruitment, all households were successful in obtaining overseas employment with earnings (about US $4500 a year) considerably higher than those in similar jobs in the Philippines (about US $800 for a full-time construction worker). The willingness to part with the most important means of agricultural production reflects the belief held by most villagers in the sample that the gains from hired overseas employment, as demonstrated by those who have returned from the Middle East, far outweigh the costs of losing productive assets like land. It also partly reflects perceptions of the prospects of agriculture relative to alternative sources of income.

The demand for land parcels which were up for sale or lease during the seven-year period came partly from Sta. Lucia residents and partly from those living in neighboring villages in the municipalities of Sta. Ana and Candaba. Outsiders, particularly those coming from the swampy lands of Candaba, prefer to have farms in the Sta. Lucia area because it is protected by a flood control dike. The outside buyers accumulated incomes from local employment, external remittances, retail stores, and agricultural production. These incomes were then invested not only in rice production but also in vegetable and poultry farms.

The local sources of demand for land in the seven-year period consisted of a particular group of villagers. Apart from the nine landless farm household heads with newly acquired 1-hectare farms and a single farmer whose landholdings increased because of inheritance and the investment of income from a small retail store, the remaining five land-gaining households had also expanded their holdings by purchasing cultivation rights in the covert land market with 'Saudi' money. At least some farmers returning from Middle East employment, then, have used part of their savings to invest in land.

The 'Saudi connection' also accounted for much of the acquisition of other productive assets. Five of the seven households in the sample which acquired Japanese irrigation pumping units in the 1977–1984 period used savings from hired overseas employment. So did three of seventeen households with new hand tractors for land preparation and the two households which acquired small mechanical threshers. The significance of the acquisition of productive assets and of savings and remittances from overseas employment becomes more apparent in the following analysis of changes in levels and sources of household income.

Changes in Levels and Sources of Household Income: 1977–1984

Comparison of the 1977 and 1984 household incomes (deflating 1984 incomes by the consumer price index) reveals that 70% of the sample households did not register significant changes in real income levels during the seven-year period. However, this stability masks changes in the relative importance of different sources of income for this group. Among farmholding households, the mean proportion of rice in total income declined from 60% to 49% while the proportions of both nonagricultural and nonrice farm incomes increased, indicating growing diversification. A number of farmers, claiming that vegetable growing and poultry raising are more profitable than rice agriculture because of the costs and risks involved in the production of the staple, transformed part of their farms into vegetable plots or small poultry sheds.

Among the landless households whose income levels remained the same, the most significant change in source of income occurred, as may be expected, for the nine new farmers; on average, 30% of their household incomes in 1984 came from rice production, reducing the proportion of hired farm employment in total income. In contrast, about 68% of the income of households without control over ricelands came from nonfarm, casual wage employment. Both the new farmers and the landless households who did not register changes in income levels during the seven-year period continued to combine agricultural and nonagricultural sources of income.

Among the remaining 30% of households with income increases of more than P10,000, more than half had members who either worked in the Middle East in the late 1970s and early 1980s or were still working there at the time of the study. Among the ten households with annual 1984 incomes of P50,000 or more, only three had no direct 'Saudi connection.' One of these moved out of rice farming and is currently drawing its income from a private business and from renting out a thresher. The other two are headed by hardworking farmers who have accumulated capital from rice production and have achieved consistently high yields except for a few cropping seasons. Both have kept abreast with the latest developments in rice production and were able to cover the initial costs of adopting the Green Revolution with interest-free loans from immediate relatives who had either overseas or local employment. Furthermore, both households have used the initial profits from rice production to diversify their sources of income. One of them invested the surplus from rice farming in the raising of pigeons from which the household obtained 66% of its total 1984 income. The other used the surplus to intensify vegetable production; by 1984, this activity provided sufficient funds to cover rice production costs.

The main difference between these two households is that the former had only one available member of working age at the time of the study. Two household members were enrolled in an expensive private school in the provincial capital; their father claimed that since he has resources to hire the services of a permanent laborer, the best legacy he can pass on to his sons is a good education. In the other household, four household members were available for farmwork since only one of four school-age members was enrolled in school.

The other households in the P50,000 and above category, which had higher incomes in 1984 because of the 'Saudi connection,' can be divided into two groups: those who returned from overseas employment before 1984 and those who were still in the Middle East or who had members abroad during the study period. The former used their savings to invest in capital goods like hand tractors or threshers, the rental of which provided around 30% of their 1984 household incomes. On the average, rice production accounted for about 37% while the share of vegetable production was about 28% for this group. On the other hand, households with members who were still in the Middle East were propelled to higher income groups because of the remittances from overseas employment. The proportion of remittances to total income was very high, ranging from 60% to 80%. These households used the extra money to maintain the same level of rice production. Two put up a small retail store while three engaged in moneylending. Key informants in the village claim that the new moneylenders are more vigilant in collecting payments.

In general, however, a large proportion of external remittances were usually allocated to purchase of durable consumer goods rather than investment while a member was still abroad. Concrete houses replaced the huts made of light materials, and consumer items like stereos and television were purchased. The proportion of children in school also increased for this group, reflecting the view that education is a mechanism for upward mobility. The returning overseas workers who invested in productive assets did so towards the end of their contracts when the prospects of renewal were dim and the possibilities of returning to the village soon became more real.

The 'Saudi connection' also accounted for the increases in income among half of the households below the P50,000 category. Except for two former landless heads who were among the new farmers in the village and one farmer who leased his land to obtain overseas employment, all of these households depended on remittances from sons in the Middle East. Some of these, however, had families of their own, so for the households in our sample the share of external remittances constituted only about 21% to 30% of their incomes. These remittances were used to cover the costs of rice and vegetable production but were not used to purchase agricultural assets.

The remaining households without remittances depended on rice earnings for about 45% to 57% of their incomes. These households also have diversified their sources of incomes during the seven-year period. Three relied on earnings from a small retail store for about 30% of their income in 1984. Aside from diversification, these households had higher incomes because compared to all farms in the sample, the yields of their 2 to 3-hectare rice farms were higher than average in 1984. Five of these households were among the few in the village with dry-season harvests. Moreover, they used to pay land rents in 1977 but none of them paid rents in 1984, increasing their share of rice output (Table 7.2).

Conclusion

In our attempt to generate ideas about causal mechanisms which propel peasant households towards particular economic and social trajectories, we cannot provide a comprehensive picture of the changes in technical conditions, the relationships among different sectors, and the overall process of surplus extraction and agrarian differentiation because of the limitations of our sample. However, a comparison of the conditions of the same peasant households through time reveals that accumulation—reflected in the acquisition of productive assets and capital goods (e.g., land, machinery, jeepneys, tricycles) and by higher incomes from diversified investments ranging from moneylending to retail stores—was made possible for a large number of households in the sample by external

TABLE 7.2. *Distribution of Rice Output by Factor and Source: Wet Crop Season, 1977 and 1984, Sta. Lucia*

	1977	1984
Internal (%)		
Hired labor	19	18
Farmer's share[a]	42	38
Barrio tax	1	1
Sub-total	62	57
External (%)		
Capital expenses	12	10
Current inputs	20	15
Rent	2	15
Interest	4	4
Sub-total	38	44
Hectarage	164.5	154.6
No. of farms	73	63
Average yield (cavans per hectare)	59	58

[a] Farmer's share is a residual which includes returns to land, own capital, and family labor.

funding which could not have been generated from intensive rice production. Those with the highest incomes and the biggest concentration of stock of means of production worked in the Middle East. Even the two 'rich farmers' without the 'Saudi connection' initially reaped the benefits of the Green Revolution because of access to external funds, part of which were also raised abroad. Those with the highest incomes also diversified their sources of income, cognizant of the fact that reliance on rice farming alone would put them in a precarious position since production costs have been increasing and natural risks cannot be controlled. It is not surprising that among those with the highest incomes, the renting out of capital goods provided greater incomes than rice production.

The prospect of accumulating in the future has made the search for overseas employment almost an obsession for some villagers in the sample. The pattern of accumulation via overseas employment may be difficult to sustain, however, given the declining demand for construction workers in the Middle East. There is also no guarantee that access to employment contracts abroad will necessarily put the household on the road to accumulation. Many households in Philippine villages have used external remittances for nonproductive ventures like constructing better-looking homes and gambling.

Aside from external remittances, the strong social organization of farmers, initially supported by external private development funds, accounted for access to particular means of production in the village. The cooperative fishpond and rice mill and the organized resistance to rental payments which has increased the farmholder's effective share of farm output attest to the fruits of organization. Among the landless, the takeover of hacienda lands was possible because of the farmers' association and village leaders who supported the takeover and the organization of the few landless.

We may then conclude that both the accelerating pace of agrarian change and the changing pattern of agrarian differentiation in Sta. Lucia in recent years have been due only in part to agrarian commercialization spurred by the Green Revolution, and much more to factors external to the village economy. In particular the export of labor to the Middle East has accounted for a large part of the changes in distribution of land and other assets and economic diversification in recent years. It is likely that similar patterns of differentiation may be found in the many other labor-exporting villages of Central and Southern Luzon.

References

Arcanes, F. (1984). *Filipino Migrant Workers in the Gulf Region: A State of the Art Report.* Tokyo: United Nations University.

Kerkvliet, B. (1977). *The Huk Rebellion: A Study of Peasant Revolt in the Philippines.* Berkeley: University of California Press.

Larkin, J. (1972). *The Pampangans: Colonial Society in a Philippine Province.* Berkeley: University of California Press.

Chapter Eight

Accumulation on the Basis of an Unprofitable Crop

BRIAN FEGAN

This study examines agrarian differentiation in a rice-cultivating village in Central Luzon in the context of the external forces discussed in Chapter 6. Through a detailed inquiry into changing patterns of access to and control over resources, we shall see how evidence of both capitalist penetration and family accumulation must be sought in indices other than operated farm size and in places beyond the village. We shall also see that when land, working capital, labor, and product have become market commodities, it is not necessarily the case that capitalist penetration will take the form of large-scale or kulak commercial farming.

In this village, a combination of government policy instruments and political forces has made rice farming so unprofitable that rural capitalists have avoided farming but instead have favored handling inputs, machine services, outputs, and credit. Even though small farmers are increasingly in debt, they are not losing their land, for the state under Marcos prevented evictions, while capitalists who do not want to farm have found new collection mechanisms.

Differentiation Within the Village

Buga is a village of San Miguel town, Bulacan province, in Central Luzon. The village has been deeply affected by the new technology in rice production and by modernization of the market, and it has excellent communications. It is 80 km from Manila, 6 km from the township, and begins 1 km upstream from busy concrete Highway 5, along a dirt and gravel village road. High school students and market women plus those with jobs or businesses in town can get there by bus, jeepney, or motor tricycle. The village has a six-grade primary school and a Catholic visita, and it has been connected to the national electricity grid since 1979. Almost all houses have had transistor radios since the late 1960s; about 25% now have television and other appliances.

The village is in a wet-rice monocrop area, and all agricultural land is planted to rice in the monsoon season (June–November), when non-rice annuals would rot. A second crop of rice or a cash crop of pickling cucumbers is planted on a smaller area, ranging from 10 to 25% of the land depending on the availability of water. Though connected to a technical irrigation scheme served by Pantabagan dam built in the 1970s with World Bank funds, the village is at the extremity of the system and cannot rely on second-crop water in most years. Some 15% of farmers own petrol or diesel pumps to raise water from river pools and tubewells in hope of a second crop.

Modern varieties of rice were introduced in 1968 with IR8, and since 1973 these hybrids have accounted for all but a tiny fraction of the planted area. The district is well serviced by chemical outlets and has been a target area for agricultural extension, land reform, and institutional credit. Standard yields rose from 2.5 tonnes/ha in 1972 to 5 tonnes/ha in 1984. Farmers have been using machines and agrochemicals for fifteen years. Of 95 households that operate some land, 38 own, or are paying off, two-wheel tractors that replaced water buffalo. Four farmers own axial-flow threshers. Four-wheel tractors custom hired from entrepreneurs along the highway and additional threshing machines enter the village in season.

I first studied the village in 1971–1973 and in restudies in 1980–1981 and 1984 took comparable censuses of agriculture, households, and persons. In the interval between my March 1973 census (Fegan 1979) and that in March 1984, rice agriculture continued to undergo intensive technical change and partial implementation of the 1972 land reform. The data focus on control of land and visible capital by 'village residents'—defined as all those who support or are supported by a household in the village but do not maintain a separate conjugal household outside of it. This definition allows inclusion of labor migrants abroad, workers

TABLE 8.1. *Operated Farm Sizes by Household: 1973 and 1984, Bo Buga*

	1973 households		1984 households	
	N	%	N	%
Riceland (ha)				
>5	6	4.8	2	1.0
4.5–4.99	9	7.2	2	1.0
4–4.49	—	—	2	1.0
3.5–3.99	17	13.6	4	2.1
3–3.49	—	—	12	6.3
2.5–2.99	17	13.6	15	7.9
2–2.49	—	—	12	6.3
1.5–1.99	3	2.4	8	4.2
1–1.49	—	—	7	3.7
0.5–0.99	1	0.8	19	10.0
<0.5	—	—	12	6.3
Total operators	53	42.4	95	50.3
Total landless	72	57.6	94	49.7
Total households	125	100	189	100

who board in Manila, etc., and return at weekends or once a month, as well as students boarding away.

The major changes merit summary before discussion to set them in historical context. The farm censuses and preliminary tabulation of household data show the following trends:

Increase in the number of households from 125 to 189.

Increase in the number of households that control some riceland from 53 (42.4%) to 95 (50.3%) (Table 8.1).

Limited emergence of illegal control of land (but not ownership or legal tenancy) by moneylenders in 1980 though this trend was reversing in 1984.

Increase in the number of landless households from 72 to 94 but a fall in the landless as a proportion of all households from 57.6 to 49.7%.

Decline in rents as a percentage of total harvest but increase in costs of machines and chemicals and in debt for them.

Increase in the number of machines, from one hand tractor in 1973 to forty in 1984, small threshers from none in 1980 to four in 1984; a steady fifteen irrigation pumps; displacement of winnowing machines by threshers.

Drop in buffalo numbers (replaced by hand tractors) with a partly compensating increase in the number of cows and goats fed on the idle grass and straw, owned by landless.

Collapse of women's pig raising due to controlled pork prices, high fodder prices, and competition by giant corporate pig factories.

Continued high out-migration of educated and skilled children of better-off families, partly compensated by in-migration of uneducated landless laborers from poorer regions, some as debt bondsmen.

Increase in house improvements, durable consumer goods, and education levels of children for prosperous farmer, employee, and remittance households, with no change for landless agricultural workers.

History of Land Control

The village was first settled in the late nineteenth century by tenant farmers taking up blocks of 2 to 6 ha on lands owned by two large absentee estates, whose owners held by the 1920s over 400 ha apiece, and by several absentee medium owners (13–50 ha) and small owners (7–12 ha). By the 1930s all arable village land was under cultivation, all of it by tenants; no villager household owned land. In the liberation period (1945–1947) several tenant farmers acquired abandoned or stolen army trucks, and from then to the mid-1950s they made money by hauling logs from the mountains. Three of these bought cheap nonirrigated land, most of it outside the village, from the heirs of medium landowners during the Huk rebellion (1946–1953). In 1963 one of these led a delegation to his namesake the President, which gained them 3-ha homesteads 40 km away in Candaba swamp, usable only in dry season and if the irrigation flows. No villager has bought land from farm profits.

The overwhelming fact about farm operators in 1971 was that they were tenants operating 2 to 4-ha farms as sharecroppers. In 1973 only 53 of 125 resident households had control of farms. The four owners and those with homesteads had two or more farms, working the main farm and homestead with a son, though two owners had tenants on outfarms. Farmholders were markedly older than the average household head. All the farmholders were tenants, on one or more farm, except for one of the owners. The standard share tenancy arrangement in 1973 was to deduct from the gross harvest the seed plus 250 kilos of paddy for reaping and 5% of the gross for threshing and divide the remainder 60% to the farmer and 40% to the owner. A few farms were held under more favorable legal sharing arrangements allowing 70% to the tenant if he bore all expenses. Despite a land reform law in 1963 that allowed tenants to sue for fixed-rent leasehold and another in October 1971 that made leasehold 'compulsory,' landowner resistance limited the number of March 1973 leaseholders to five. Most landowners charged 50% interest for both production and consumption loans during the work season, payable at threshing. One of the two large estates was run on an incentive system; it charged only 8% interest on production loans but demanded the prevailing rural rates of 50% for rice loans and 10% per month on cash consumption loans.

When the 'land-to-the-tiller' program was decreed in 1972, landowner resistance first took the form of removing land from the reform by dividing title, removing tenants, changing crops, etc. Then they turned to forcing up the lease rentals or valuation by noncooperation, delay, obstruction, legal challenge, filing harassing charges, and attempting eviction. The object of these maneuvers was to raise land valuation and lease rents in order to capture part of the windfall benefits from the higher harvests and the chance of more crops per year that flowed from the Green Revolution and massive government investment in irrigation. By about 1978 valuations and rents were fixed and only the diehard large landlords of San Miguel persisted in organized resistance. Landowners ceased then to use overseers to police the harvest, as the amount of harvest became irrelevant once rents were set. Despite corruption, muddle, incompleteness, and continuing landowner resistance, the land reform broke up the large estates as institutions. Tenants then refused to use estate threshing machines and tractors, which landowners had to sell or hire out in competitive markets.

Landowners

By the time of the 1984 restudy, landownership and control had become more dispersed. To understand this it is necessary to take separately the owners, the tenants of 'legal farms,' and the operators of actual farms. Among the landowners, ownership appeared more dispersed both as a result of real division among heirs and of manipulations of title in 1973–1974 to avoid land transfer by making title appear divided among several persons, each with less than the limit. However, the meaning of land 'ownership' had been weakened as owners had lost the right to remove tenants even in the event of repeated nonpayment or underpayment of land amortization or lease rent. Although most owners had managed by prolonged resistance to force a higher valuation or lease rent than that based on pre-1972 harvests, landowners lost any share in the harvest increases from new technology and irrigation that took place after about 1976. Where the standard yield of irrigated land in the wet season in 1972–1973 was 2.5 tonnes, by 1980 the standard was 5 tonnes. Moreover both CLT (Certificate of Land Transfer) holders and tenants had large arrears of unpaid amortization and lease rent from both nonpayment and partial payment. Landowners had lost political and social power over tenants with the loss of the sanctions of eviction, control of credit, and a relative monopoly of machine services and chemical supply. Tenants no longer behaved deferentially.

Large landowners lived in Manila with urban investments and professions. None had entered agribusiness in the region either as farmers or as suppliers of inputs, machines, and credit. One, Soriano, who ran a rural bank from about 1970, lost money and in about 1979 sold out. Several small and medium landowners of the active town-resident bourgeoisie had diversified their investments into transport, light engineering, contracting, grain handling, and raising pigs and poultry commercially. Some with commercial piggeries of below 500 sows had closed by 1984.

Small and medium landowners with under 24 ha live in the municipal township. A handful managed to evict or buy out some or all of their tenants in anticipation of the 1972 land reform. However, they complained in 1984 that farming was unprofitable. Alberto Solomon, owning 35 ha, (10 ha without tenants), ran an integrated agriservices enterprise. He claimed that the key to good business in the 1980s was to supply machine services and chemicals on credit, on condition that the farmer use the lender's threshing machine to ensure collection. Solomon claimed that this investment in tubewells for his 10-ha wage-labor rice farm would have been lost had he not turned to speculative vegetable crops in the dry season, worked by petty lessees.

Farm Operators

From the tenant's point of view, land reform despite its limited implementation provided above all security of tenure and lower rents as a proportion of yield. Although the operator now bears all costs, the price of rice is kept low by government policy. The nonrent costs of production have risen, leaving a thin margin in a good year. On risky land—i.e., rain-fed wet riceland along the foothills and low-lying land on the fringe of Candaba swamp prone to flooding—many farms were not worked in the 1983 wet season because the operators had in effect become bankrupt. However, these farms were not being taken over by creditors because of low and risky returns. In effect marginal land was being abandoned in face of unfavorable prices. On irrigated land with relatively low flood risk, no land was abandoned, but many farmers were in debt arrears for unpaid amortization of lease rent, machines, and expendable inputs credit, while yields fell in 1983 as they reduced fertilizer in response to its high national price following peso devaluation and high interest rates. However, institutional creditors and landowners lacked effective means of enforcing payment. The government, though it had lost the will to press through the land reform, lacked the political will to collect debts. Institutional credit for crop expenses dried up from the late 1970s, and by 1983–1984 main-crop credit was available to only six of the ninety-five farm operators in the village.

In the wake of land reform the meaning of 'farm' has changed. The 'legal farm' is the subject of relations between landowner and tenant controlled by the Ministry of Agrarian Reform. By law, a legal farm held under CLT is not divisible or transferable except to a sole heir. However, several whole farms have been permanently transferred since the reform to a person who purchased the right from the tenant, often after an initial loan in return for the usufruct. Then with collusion of the outgoing tenant, village headman, landowner, and ministry officials, the legal farm right was transferred. More to the point, many legal farms have been divided among heirs of a retired or deceased legal tenant, or usufruct of part of a legal farm has been surrendered to a moneylender as pawn for an unpaid loan. Where there were 53/125 households in 1973 who operated farms, division of farms among heirs and by pawn made 95/189 households operators in 1984.

However, this apparently more egalitarian distribution of land masks a polarization in control of operated area. The better-off farmers manage to set aside an 'establishment fund' (Fegan 1979, 1983) sufficient to educate their children for urban white-collar jobs, largely by enforcing a delayed pooling within the household. In effect, after the family has funded the education of one child, he or she is expected to delay marriage until remittances added to the parental household income have funded the next, and so on. In such families only one or two children remain in the village, one taking over the undivided farm as his or her portion. Families that do not pursue or succeed in this strategy may be left with several children who at marriage have no secure means of living. When they divide the legal farm, the operated units given to each child run from 0.5 to about 1 ha, which at current rice prices is below the minimum viable area of 1.5 ha in irrigated one-crop land. After meeting costs of production the net harvest is not sufficient to maintain a family, and since the operated unit is not a legal farm, the operator has no security of tenure or entitlement to cheap institutional credit for machines, cash, or inputs. It is among the 31 households with under 1 ha (16.3% of all households and 32.6% of operator households) that the highest rates of interest are paid. Few own machines. Should the strategy of rural capitalists switch back from lending against the security of the crop tied to the threshing right to lending against the land as security—then it is probable that much of this land would be reaccumulated by moneylenders, albeit in scattered plots.

The process of dispersion and division of legal farms among heirs has been unusually rapid in this village because in 1973 the number and proportion of aged farmers was high (Fegan 1979:figure 13). In nearby villages the other process of division of farms by pawn to moneylenders went ahead in the 1970s, but it had come to a halt by 1984. After 1979, lenders I spoke to found it less profitable to gain control of farms than to gain control of harvests as loan security using the small thresher to collect. It is instructive to examine their operations.

Rural Capitalists

The major moneylender in adjacent Bo Cambio controlled in 1980 some nine farms averaging 3 ha apiece of irrigated land, placing the borrower operator under a covert share tenancy arrangement while the lender operated an integrated services business. The lender Vito supplied certified seed from his own licensed seed-producer's farm (at double the ordinary price), fertilizer, chemicals, and services of his hand tractor, plus cash for transplanting expenses, all at retail rates plus 10% per month interest. The borrower-operator was obliged to work under supervision and 'learn both good farm practices and financial management.' He contracted to use the lender's thresher. From the gross harvest they deducted costs of these inputs and services plus the cost of reaping, threshing, irrigation and land rent and then divided the net fifty-fifty between farmer and creditor. As well as these integrated services supplied to borrowers who had pawned their usufruct right, Vito also supplied any element of the package at 10% per month interest to other borrowers, on condition the borrower use his thresher.

By operating across linked markets for several inputs, and with an inescapable collection mechanism in the thresher, Vito made higher profit than in single markets, because of his contractual monopsony. At harvest Vito's thresher, run by a son, collected debts plus interest and its own 6% of the gross, with priority over any other charge on the harvest, allowing him to lend without risk. This business allowed him to buy fertilizer and chemicals for himself and his clients in bulk while charging it out at the prevailing retail price. He secured continuous work for his hand tractor and thresher in an increasingly competitive market. At the threshing floor he collected rice at the prevailing low harvest price, but he then held the rice in his 100-ton warehouse awaiting the postharvest rise. Meantime, from the house his wife lent cash to other women for living expenses and trading capital, at the prevailing 10% per month interest, and in season operated as a buying station for rice, turning over capital quickly at low margins. Vito has other sidelines as a commission agent for farm machines, for which he can arrange bank credit. He is a past national vice-president of the Land Reform Farmers Association with a long history of fighting for tenants' rights against the landlords. By 1984 Vito had ceased to lend against usufruct, had withdrawn farm control, and operated an integrated services and trading business.

In Buga, small-scale lenders operate similar integrated services businesses. The most prominent in 1984 was Totoy, graduate of the Agricultural College at Munoz and employed as an extension agent for farm credit by the Rural Bank of a town some 40 km away. Totoy's father holds a 4-ha farm in Buga, but Totoy lives in a subdivision on the edge of the township and commutes between home, job, and borrowers by motorcycle. With less capital, Totoy could not supply the whole of Vito's integrated services, but in 1979 he began lending money for fertilizer as institutional credit dried up. At first he lent at 10% per month interest against the security of usufruct of part of the farm should the borrower default. However, this proved a bad investment as he had no effective collection mechanism until in 1982 his wife's brother Siso bought a threshing machine on Totoy's advice. They began operating together, and Totoy lent money for fertilizer on condition that Siso's machine did the threshing. In 1983 they added land preparation, with Totoy's brother Marte or Siso supplying plowing services, Totoy supplying fertilizer, and all collecting through Siso's thresher.

Similarly, Pablo with CLT on a poor rain-fed farm bought first one thresher in 1982 on bank loan, then another in 1983 'since business was so good', and mounted one on a jeep. Faced by competition from thresher operators who could supply fertilizer credit, Pablo linked up with a fertilizer outlet in town that faced a drop in sales because credit had dried up but lacked a collection mechanism. With Pablo guaranteeing the loan, and taking a 2% commission for collecting it, he was able to link farmer and chemical distributor while securing business for his machine locally. His son or nephew accompanies his machines to distant towns in the harvest season where commission agents line up whole village sections on threshing-only contracts.

In the highway barrio of Ilog Bulo, two women capitalists who began as rice traders added small mills and then entered integrated services businesses by diversifying into supplying fertilizer on credit plus four-wheel tractor and threshing machine services. Though the scale of operations was larger, the key was to remain separate from farming itself but supply inputs and services on credit, collect through the threshing machine, and trade in rice. Tracing inputs, machine services, and credit links beyond the village showed that the integrated services capitalist enterprises took off after 1979, coinciding with the breakdown in institutional credit and appearance of the light axial-flow threshing machine designed by IRRI.

So long as moneylenders had no security for the loan aside from taking over part of the farm in case of default at the harvest, rural capitalists were unwilling to lend except to neighbors about whom lenders had adequate credit ratings and over whom as fellow villagers they had sufficient sanctions to seize usufruct of the pawned land in event of default. Operating scattered small plots of land was unattractive. In practice lenders left the defaulting farmer as a kind of tenant, taking 50% of the net crop like old landlords. The threshing machine collection mechanism cut this restriction. By allowing for a straight services contract with secure collection beyond the lender's village, it opened the way to rural capital.

In early 1984 I was unable to discover a case of a borrower breaching contract to give the lender's machine the exclusive right to thresh. It is notable that in the old landlord system the first farm operation mechanized was threshing from about 1918, when landowners introduced steam-powered machines to collect rent and debt. However, the heavy McCormick-Deering threshers used from about 1939 until the 1970s could be hauled only on dry roads and fields and handle only dry straw. The IRRI-designed light axial-flow threshers can operate on wet-season roads and be towed to each paddy even through mud. About one in three in 1984 had been modified in village workshops to mount them on an old jeep body increasing mobility and hence earning power, and to take a diesel motor and hence lower fuel costs.

This style of rural capitalism, standing back from an entrepreneurial role in production itself, is a variation on the dominant merchant or rent-capitalist style of operation (Fegan 1981b). This strategy entails avoiding direct involvement in production, avoiding outlaying capital at risk in wages and materials, but drawing to the capitalist a number of dependent petty entrepreneurs who use their own labor and drive that of others in the production process while bearing risk. The capitalist takes a merchant profit on inputs and outputs, lends money at interest, and lets out units of durable capital like land or machines in separate units operated by dependent petty entrepreneurs. All the machine owners discussed above put machines in the hands of a close kinsman or trusted dependent petty entrepreneur, who operated it under a piecework or output-share system. This was a per hectare fee for four-wheel or two-wheel tractors; for threshers it was 0.5% of the grain threshed, taken from the machine owner's 6%. The effect of the nonwage system of labor payment was to force these operators to act as dependent petty entrepreneurs. They acted entrepreneurially to find contracts for the machine, worked fast and well for long hours so as to maximize their earnings, conserved fuel, moved the machine quickly between jobs, and kept it in repair. This system needs no supervision by the machine owner and is remarkably well adapted to allowing the owner to stay at the warehouse to make buy-and-sell deals in other parts of the diversified enterprise. The general features of rent capitalism were first noted by Bobek (1962), and their modern variants in the Philippines were analyzed in an earlier work of the present author (Fegan 1981). Marxists may feel more comfortable in regarding this as a special if common form of merchant capital paralleled by the rent or profit-share relation between the operator and the owner of a taxicab or a business franchise.

The Landless

The landless households increased from seventy-two in 1973 to ninety-four in 1984, an absolute increase of twenty-two households. But they declined as a proportion of all households from 57.6% to 49.7% in the same period as older farmers retired or died, multiple farms were dispersed and single legal farms divided, among their children. Where only one household in 1973 held a farm under 1.5 ha, by 1984 there were thirty-eight households operating nonlegal farms under this minimum viable area. Households with less than 1.5 ha could not live on the net farm produce at current prices without supplementary income for the head or other members. A variable number of households, most of them landless or land short, rent irrigated land in dry seasons from farm operators to grow labor-intensive crops of cucumbers. These are sold through agents to a pickling factory in Manila.

However, land is not the only or most important capital of village households. Other important productive means include agricultural machines, livestock, jeeps and tricycles plying for hire, sewing machines, moneylending capital, shops and trading enterprises, and mechanic's, carpenter's, and beautician's equipment. In addition many villagers have invisible capital in the form of formal educational qualifications for government salaried jobs or skills and contracts as tradesmen, drivers, construction contractors, bulldozer operators, casino croupiers, movie technicians, and so on. Although the household and personal census forms for 1984 have not all been analyzed, a sample indicates that the 1973 pattern persists (Fegan 1979:ch. 6) in which both individuals and households pursue diversified sources of incomes, using their own or rented and borrowed capital, their skills and contacts, to find business or labor openings in order to subsist and perhaps even prosper. Between 1973 and 1984, however, the prospects for those with limited education and no skills beyond agricultural work had worsened because the agricultural real wage had fallen against the CPI, while agricultural labor had been displaced by machines and chemicals in the second wave of the new technology from about 1978 (Jayasuriya 1984). Locally, the cucumber crop, planted by landless, failed in 1984 when water shortage hit the crop, completing destruction caused by newly insecticide-resistant thrip insects. Many urban workers in factories and services in Manila and the township lost their jobs in the economic downturn precipitated by international debt problems and assasination of Senator Aquino.

The landless and land-short households need to be stratified according to whether they are dependent on farmwork or have substantial capital and nonfarm income before meaningful statistical statements can be made about differentiation. A preliminary tabulation shows that thirteen of seventeen male labor migrants to OPEC countries, the U.S., and Australia and both workers on U.S. bases are landless. Their remittances are used not only for house improvements and consumer goods but investments like jeeps, tricycles, a threshing machine, buying farm rights, educating children, and above all trading and moneylenders' capital used by their wives. Other landless and land-short households have members with secure salaried jobs as primary and high school teachers (7), agricultural extension agents (2), bus inspectors (3), bus drivers and conductors (4), electricians (1), bulldozer driver (1), police (1), bodyguard (1), trader (2), stockraiser and moneylender (2). It is common for salaried women and wives of salaried men to trade or lend money with the twice-monthly pay.

It is landless women who have lost most rural work since 1973, without compensating new rural opportunities to earn. In rice agriculture women have been technically displaced from a number of jobs, beginning in the mid-1960s. The petrol-driven winnowing machine quickly displaced women from a legitimate 2% of the threshed rice in the mid-1960s plus at least another 1/2% in covert earnings by concealing rice in the straw and through the farm wife's generosity to kin and neighbors which allowed generous measure for the task. Women rarely engaged in the harvest tasks of reaping, hauling sheaves, and threshing in the early 1970s for the main crop, but poor women reaped in the urgent wet-season harvest. They also gleaned, from the stubble, by rethreshing hand-threshed straw and sifting straw threshed by the large McCormick-Deering machines. By 1984 new varieties and the efficient small thresher had left little to glean. Women also retreated, or were pushed out, from the reaping when the IRRI threshing machine broke the labor bottleneck in the wet-season harvest. Women had always dominated transplanting and hand weeding. They gained an initial increase in the transplanting labor with the introduction of straight-rowed planting in the late 1960s. But rows permitted the rotary push weeder, classed as man's work. By the end of the 1970s imported chemical weed-killers improved in effectiveness and fell in price faster than the falling real rural wage rate (Jayasuriya 1984; Coxhead 1984), leading farmers to reduce both transplanting and weeding labor. Farmers in Central Luzon found they could control weeds by chemicals and shifted to broadcasting germinated seed direct onto the paddyfield mud (direct wet seeding, DWS), displacing transplanters from almost all the dry-season crop and some of the wet season. Weed control by chemicals displaced female transplanting and hand weeding plus male labor in uprooting seedlings and rotary push weeding. The heads of labor contractors for women transplanting gangs said that there is only about half the former transplanting work. It is possible that recent devaluations of the peso, shortage of foreign currency for imports, removal of credit subsidies, and rural credit shortage will reverse this trend in which socially scarce capital was replacing plentiful and cheap labor while state-subsidized credit created an artificially low private cost of capital (David 1983).

But one of the most important losses to women was raising pigs as a household sideline or small-scale commercial operation. Until the early 1970s the small black native pig was tethered in the houseyards and fed bran returned by the miller, food scraps, forage, and feces. From about 1970 larger European breeds were introduced, housed in concrete sties in the yard, and fed commercial hogmash. Relative prices in 1973 made this a profitable enterprise, compatible with running a household. At that time thirty-one of fifty-three farm households, and forty-four of eight-five landless households, had one or more pigs. By 1984 most of the sties were empty, devaluation of the peso translated world dollar prices of soy and maize for fodder into rising input prices, while the government attempted to hold down the price of pork. The result favored large commercial pig farms whose operations integrated feed mills, pig raising, and in some cases feed grain farming. In the mid-1970s a pollution zoning law forbade pig farming within fifty miles of Manila, but San Miguel stood just outside the zone. Five corporations set up giant integrated pig factories with 10,000 to 20,000 breeding sows apiece and a monthly turnout of 1000 to 2000 marketable pigs. The corporations were losing money in early 1984 but claimed reserves sufficient to hold on until the government allowed legal pork prices to rise or until black market channels could grow. Small operators could not persist at the prices, cutting out thousands of women from an important income source.

No village men or women had taken jobs in the pig factories, where wages are low and employees are quarantined within the patrolled compound for four weeks continuously before taking a week's leave. The Chinese owner of one factory enunciated a clear policy of drawing labor from the poorer islands and mixing all work teams ethnically to minimize the possibility of a strike. A sample of household and personal protocols shows some young women found piecework in textile factories in Manila, but none had entered the export processing zone in Mariveles. The Manila workers board, returning once a month. Piecework rates are low and there is no security; when the employer lacks orders or materials, the girls are laid off.

Mechanization has reduced the amount of work for male landless agricultural workers and made some parts of it dangerous. Most male-reserved tasks are heavier, more skilled, sometimes dangerous, and often individual work requiring judgment rather than repetitive work in a team as with transplanting. The number of working days per hectare or per ton of output has been reduced by the new technology, but the rise in labor productivity has not meant a rise in real wages per day or task, and workers say the number of days available per year has fallen. This last is difficult to verify on recall interviews alone. I have not been present through a main crop since 1972 and did not see through the 1981 or 1984 dry-season crops.

The hand tractor displaced mainly the farmer's and his son's unpaid labor in land preparation with buffalo and the daily chores of tending the beasts year-round. Only sixteen of seventy-two landless households owned buffalo in 1973, but forty-seven of the fifty-three farm households. In 1984 some six landless men kept buffalo, which they hire out to plow the edge of the bunds missed by machines. Others had switched to raising Indian cows and goats. From the late 1970s weed-killers displaced rotary push weeding. By 1983 the task of uprooting seedlings had been virtually eliminated for the dry-season crop and reduced for the wet season by weed-killers making possible direct wet seeding.

After crop establishment, the managerial tasks of spreading fertilizer and applying insecticides and weed-killers are skilled tasks formerly the job of the farmer. However, most farmers prefer to hire day labor or a farmhand to do jobs involving chemicals after several men were made gravely ill. Doctors at the emergency hospital in the township say that nerve poison insecticides are the main problems; victims usually recover after shots of atropine, but several have died on the way. Chemical companies and distributors still do not issue large-print safety instructions in the local language; fine-print warnings in a foreign language are of little effect on a middle-aged peasant without glasses. Peasants do not invest in protective clothing, goggles, and respirator, and it is doubtful if workers can be persuaded to endure unventilated gear in the tropical summer.

That holders of legal farms and those with adequate household income from other sources continue to fund high school and nowadays college education means that sons who might otherwise contribute to unpaid family labor on the farm are in school, have graduated and found work outside the village, or are too proud to labor. This leaves aging farmers unable to handle the heavy land preparation and unwilling to spray and broadcast chemicals. Changes in household labor availability and dangerous chemicals have opened jobs for agricultural laborers, either as a farmhand for the whole season or as day or task-paid labor.

Farm operators who are old or have nonfarm jobs prefer to have a fulltime resident farmhand who can be directed to do any work, including chores in the houseyard. Village landless rural workers are unwilling to accept this role for a fixed per season fee payable at harvest, because they need the cash or food in the preharvest hunger period, because the recent rate is too low, and because the additional tasks prevent them taking casual day work if it becomes available and reduce them to the status of errand boys. Into this slot farmers have introduced fulltime farmhands migrating from poorer regions and islands, principally Bohol, Cebu, Bikol, and Samar. Some of the last in 1981–1982 were refugees from NPA–army clashes. Generically referred to as 'Bisaya,' these uneducated and low-skilled men came into the region initially as migrants and debt bondsmen to work in rice warehouses humping 50-kilo sacks and raking rice drying in the sun. In 1980 some older farmers without available sons 'ransomed' debt bondsmen from agents in Manila and then cut out the debt by deducting nominal wages equivalent to about one-third the going rural wage rate. The Bisaya are lodged under the farmer's house, in a disused buffalo shed, or allowed to build themselves a shack with bamboo, grass, and scrap iron if they have a wife or if several want to live together. Contracts vary but run to about 0.75 tons for a 3-ha farm, if the farmhand provides his own food. A farmhand may be allowed to work for wages for others on his own account if the farmer has no work for him, but he must work whenever directed. The Bisaya are desperately poor, their total possessions a change of clothes, a ricepot, and a blanket. Although they are used as ratebusters by the farmers, the local landless agricultural workers express pity for them rather than anger. However, several beatings after drinking parties had got out of hand belie this.

The threshing machine reduced the amount of labor in the harvest by about half, because as well as threshing and winnowing it is able to move quickly from site to site, cutting out the need to bundle, haul, and stack the reaped stalks into stooks of sheaves. This broke the harvest bottleneck caused by farmers refusing to use the big McCormick-Deering machines of the landlords around 1978. The harvest rate for *all* tasks was one-fifth or one-sixth of the crop under the old system, depending on whether the farmer contracted separately to haul in the sacks. With the new machines the reaping rate was first reduced to one-ninth. In the 1980–1981 main crop farmers tried to reduce this to one-tenth, but a typhoon and flood caused a bottleneck that year that allowed laborers to hold the rate. While there were insufficient threshing machines, harvesters could simply be 'too busy on prior contracts' to reap and thresh a ratebusting farmer's flood-damaged crop. By the 1983–1984 main season the thresher had broken the bottleneck and with it labor's bargaining power. The rate fell to one-tenth, and in early 1984 there was talk of one-eleventh in other barrios. Whatever might be argued by IRRI about the advantages of shortening turnaround time and the possibility of multiple cropping, in areas of single-crop water supply the thresher caused a large gain to capital but a large loss to labor. Harvest labor once earned one-sixth (16.6%) and now earns one-tenth (10%); the threshing machine earns 6% of which 1% goes to its crew and the rest to fuel and to capital. This 5.6% of the harvest lost to the laborers is 5.6/16.6% of their previous harvest share, or a loss of 33.7% to harvest labor as a whole. Some harvesters argue that the machine has reduced the hardship of their work by taking out the heavy threshing and hauling, leaving them free to maximize the area they reap. This cannot work for labor as a whole, unless there has been some reduction in the number offering for reaping or a compensating yield increase. There is no evidence that either has occurred; in fact there is evidence to the contrary (Jayasuriya 1984).

Conclusion

In the field area, a conservative land reform was legislated and began to be implemented just when the new technology was about to take off. It had among its objectives preservation of the family-size farm in tenanted rice and corn land by securing tenure, lowering rents, and making inputs available through subsidizing credit so that the production possibilities of the new technology might be obtained. Thus it could be argued that the Lenin-Kautsky predictions would have worked out but for state intervention designed to preserve the smallholder in order to quell rural unrest. Next, it could be argued that the peculiar cultivation requirements of rice are not amenable to large units with machines and hired labor, particularly where low-level peasant resistance poses problems for labor organization and discipline so that large operated units would attain lower yields than smallholders able to fine-tune crop care on small areas through a combination of Chayanovian self-exploitation and close management motivated by incentive.

Although these may have played a role, the most important part of the explanation is to be found outside the village: rice production under the terms of trade and natural risks prevailing in the Philippines in the last decade has been unprofitable. Mobile capital has sought other investments, and only those village-based entrepreneurs who lack knowledge of or access to other investments have engaged their money in farming.

In an environment of an unprofitable crop, and a set of state policy instruments that intervened in markets for technical inputs (subsidized credit, dear fertilizer, fuel, cheap machines, and chemicals), capitalist entrepreneurs stay out of farming finding niches upstream and downstream of the farm. Within the village, population increase after the end of the land frontier in the 1930s created a landless class. The 1972 land reform gave tenants security of tenure, cheapened rents as a proportion of the crop, and broke the power of the landowners to prevent subdivision of farms. Although it is illegal, this has allowed the increasing population to share access to land more widely, but that means only that the children of deceased or retired legal farmers, not the already landless, may get a small area to operate. In effect, the children of the most prosperous villagers (be they farmers or landless) continue to be educated or supplied with capital to leave the village, through a process of delayed family pooling that I have called the 'establishment fund.' Growing numbers of landless agricultural workers have lost access to employment as the new technology reduced man-days per hectare per crop in the second phase of the new technology, and agricultural wages have fallen against the CPI but only slightly against rice (because that has been falling against the CPI). The landless agricultural workers have also become internally stratified as uneducated and low-skilled immigrants from poorer regions compete, particularly in the job slot of farmhand opened by the education of farmers' sons. Women have lost most in labor incomes and are now largely irrelevant to the production of the crop, while sideline agricultural incomes from raising pigs and chickens have become unprofitable under 1984 prices and competition from corporate pig and poultry farms.

The new technology, under the conditions noted above, has preserved the merchant-creditor-rent-capitalist style of operation in which the capitalist breaks up productive equipment into units that are let out under the management of a dependent petty entrepreneur, while avoiding direct investment in wages and materials of production unless these can be made a debt of a small operator, risk shifted to him, and interest charged. This strategy allows the maximum proportion of capital to be kept free for merchant dealing when the seasonal level of business peaks, and the capitalist to be free of time-consuming supervision of labor, which is made to self-supervise through the necessity to make a margin over the rent and interest owed to the capitalist.

References

Almeda, J. P., and E. C. Gonzales (1984). "Daily Wage of Farm Workers in the Philippines: Crop Years 1975–1983." Economics Research Report. Quezon City: Ministry of Agriculture, Bureau of Agricultural Economics.

Bobek, H. (1962). "The Main Stages in Socio-Economic Evolution from a Geographical Point of View." In P. Wagner and M. Mikesell, eds., *Readings in Cultural Geography*. Chicago: University of Chicago Press.

Coxhead, I. (1984). "The Economics of Wet Seeding:Inducements to and Consequences of Some Recent Changes in Philippine Rice Cultivation." Masters dissertation, Australian National University.

David, C. C. (1983). "Economic Policies and Philippine Agriculture." Working Paper 83-02. Manila: Philippine Institute for Development Studies.

David, C. C., and A. M. Balisacan (1981). "An Analysis of Fertilizer Policies in the Philippines." Workshop on the Redirection of Fertilizer Research, Tropical Palace, Manila, 26–28 October.

Fegan, B. (1979). "Folk-capitalism: Economic Strategies of Peasants in a Philippines Wet Rice Village." Ph.D. dissertation, Yale University.

——— (1981). *Rent-capitalism in the Philippines*. The Philippines in the Third World Papers, Series no. 25. Quezon City: Third World Studies Center, University of Philippines.

——— (1983). "Establishment Fund, Population Increase and Changing Class Structure in Central Luzon." *Philippine Sociological Review* 31: 31–43.

Jayasuriya, S. K. (1984). "The Philippine Rice Economy: Some Notes." Seminar paper, Department of Economics, Research School of Pacific Studies, Australian National University, Canberra.

Jayasuriya, S. K., and R. T. Shand (1983). *Technical Change and Labor Absorption in Asian Agriculture: An Assessment.* Occasional Paper no. 35. Canberra: Development Studies Center, Australian National University.

Unnevehr, L. J. (1982). "The Impact of Philippine Government Intervention in Rice Markets." Paper presented at the IFPRI Rice Policy Workshop, Jakarta, 17–20 August.

Part Four

Malaysia

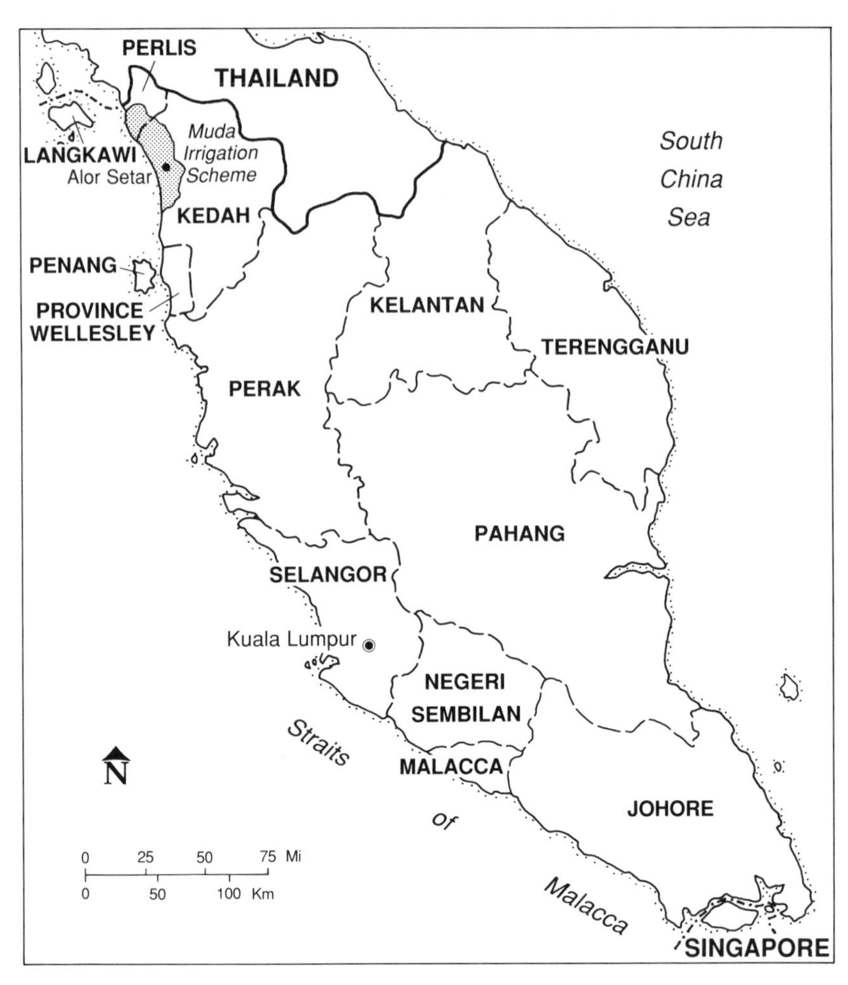

Map 3. West Malaysia

Chapter Nine

Malaysia: Rice Peasants and Political Priorities in an Economy Undergoing Restructuring

LIM TECK GHEE AND
MUHAMMAD IKMAL SAID

Malaysia is often regarded as a successful example of economic development in the Third World. Well endowed with natural resources, the country has enjoyed strong growth rates in the past two decades (real GDP went up by an estimated 6% in the 1960's to 8% per annum in the 1970's), and with a per capita GNP today estimated at US$1994 the country is more comparable in its level of economic and social well-being to the range of middle-income countries of the world rather than to most other Southeast Asian countries.

Indeed, important changes have occurred in the structure of the economy since 1957 when independence from British rule was granted. A colonial economy heavily dependent on the export of tin and rubber has been transformed into one that is not only more diversified but also more urban and industrial-based. Especially impressive has been the growth in manufacturing. In 1980, manufactured goods accounted for 30% of total export value and provided a total of 803,000 jobs or 15.8% of the country's total employment (Malaysia 1981:81).

Despite these developments, the importance of agriculture is still relatively high. Agriculture (including fishing and forestry) continues to be the largest contributor to the country's revenue and absorbed four-tenths of the working population in 1980. Although the proportion of the labor force engaged in agriculture has been reduced considerably in the last two decades from 50.5% of the workforce in 1970 to 39.7% in 1980 and an estimated 35.5% in 1985 (Malaysia 1984:127), the number of agricultural workers has continued to grow in absolute terms and now stands at 1.98 million compared with 1.71 million workers in 1970.

The agricultural economy of Malaysia can be divided into two categories: one is the plantation sector (almost entirely export-oriented) with a considerable but declining element of foreign ownership and almost entirely concerned with perennial tree crops (notably rubber and oil palm); the other is the peasant sector of export-oriented smallholders and food producers growing for home consumption or sale to domestic consumers. While the plantation sector is well organized and enjoys high levels of productivity, it has also for many years been the recipient of government largesse, in the form of access to the most productive or best-positioned land, infrastructural support, and state-sponsored technological improvements. In contrast, the peasant sector (with the exception of the larger farming units in public land development schemes that are highly organized and can attain as high levels of production as plantations) has in the past suffered from neglect or has been discriminated against by government policies (especially during the colonial period)[1] and is beset by some of the economic problems that plague small, unorganized producers elsewhere in the Third World. At the same time, there is a small population and ample land. (By the standards of most Asian countries Malaysia has a remarkably low population density; its 202 people per sq km of arable land contrasts starkly with Java's 1100 people.) This has meant that the peasantry has been largely spared the problems of minuscule farms, landlessness, and extreme exploitation by landlords that plague many other agrarian communities.

Within the peasant sector, specialized paddy farmers constitute the second largest group (next only to rubber smallholders) and in West Malaysia comprised 150,000 households or 15.7% of the total number of households in agriculture working a total of 764,160 hectares in 1980 (see Table 9.1). However, these figures underestimate their number and farm area since many farmers listed as tree-crop smallholders also plant some paddy as a subsidiary crop. Paddy farmers are to be found in all the states of Malaysia and range from slash and burn shifting cultivators producing entirely for home consumption to highly commercialized units utilizing sophisticated drainage and irrigation facilities and other modern technological inputs.

TABLE 9.1. *Number of Households by Sector in Agriculture: West Malaysia, 1980*

Agricultural sector	Total households (thousands)	Share of total (%)
Rubber smallholders	422	44.2
Oil palm smallholders	25	2.6
Coconut smallholders	34	3.5
Paddy farmers	150	15.7
Other agriculture	172	17.9
Fishermen	43	4.5
Estate workers	112	11.7
Total	958	100.1

Source: Malaysia (1976:163).

However, they are most numerous in the northern states of Kedah, Perlis, Perak, Kelantan, and Terengganu where large-scale government-initiated irrigation and development schemes, especially since the late 1960's, have converted large areas from single to double-cropping and raised yields substantially. In 1983–1984, peninsular Malaysia produced about 1.36 million tons of paddy, of which about 87% was derived from irrigated land and 13% from nonirrigated rain-fed land. Less than 1% of Malaysia's paddy production came from dry paddy. (See Table 9.2 for figures of paddy area, output, and yield.)

As a group, paddy farmers in Malaysia are important not only because they produce the major food item in Malaysian diets[2] but also because, being almost entirely Malay, they comprise a significant political element in this multiracial society where ethnic politics provides the main source of electoral support for the contending political parties. The fierce competition for the political allegiance of the rural Malay population which has continued unabated until today has meant that the problems of the rural community are often in the public limelight, and it also accounts for the many policy initiatives undertaken by the ruling government to improve socioeconomic conditions in the countryside. When general elections were held throughout the country for the first time in July 1955, it was necessary for UMNO,[3] challenged by a rival Malay party, the Pan Malayan Islamic party (PMIP), to devote some attention to alleviating the depressed conditions of the peasantry. PMIP victories in the economically backward states of Kelantan and Terengganu in the

TABLE 9.2. *Average Annual Area Planted and Paddy Yield: West Malaysia, 1950–1983*

Period	Main-season wet paddy		Off-season		Dry paddy		Total		
	Area ('000 ha)	*Paddy yield (t/ha)*	*Area ('000 ha)*	*Paddy yield (t/ha)*	*Area ('000 ha)*	*Paddy yield (t/ha)*	*Area ('000 ha)*	*Total output ('000 t)*	*Paddy yield (t/ha)*
1950–55	269.7	1.92	2.7	1.72	13.7	0.8	286.1	534.5	1.0
1955–60	286.5	2.09	3.3	2.08	18.4	1.3	308.2	629.4	2.0
1960–65	346.2	2.46	21.4	2.68	21.1	1.3	388.8	936.8	2.4
1965–70	369.2	2.52	85.0	2.85	21.8	1.2	476.0	1197.6	2.5
1970–75	369.5	2.72	199.9	3.15	12.5	1.3	582.0	1649.8	2.8
1975–80	335.8	2.87	193.1	3.36	8.13	1.2	537.1	1622.2	3.0
1980–83	295.3	3.25	194.9	3.29	6.31	1.3	496.5	1680.3	3.4

Sources: 1950–1976 (Taylor 1981:53); 1977–1983 (Kementerian Pertanian 1985).

1959 general elections as well as the redelineation in 1974 (and more lately in 1985) of parliamentary and state constituency boundaries which further favored rural (i.e., Malay) representation brought home the need for agricultural policies that were more responsive to the farming community.

In 1970 the Malaysian government adopted the New Economic Policy (NEP), a twenty-year program aimed at poverty eradication and socioeconomic restructuring to improve the position of the Malay community vis-à-vis the non-Malays. The implementation of the NEP has generated a steady stream of development initiatives directed towards the rural population.

The paddy sector, in particular, has benefited from these state policies, which, besides the massive public investment in drainage and irrigation, include restrictions on imported rice, generous fertilizer and price subsidies, and other forms of state intervention in the market to benefit producers.[4] Thus public expenditure on drainage and irrigation was increased from M$38.3 million in the 1956–1960 period to M$328.6 million in the 1966–1970 period (US$-M$2.4). This represented 16.8 and 36.0% of the total actual public expenditure on agriculture in the two periods respectively. Public expenditure on drainage and irrigation has been kept at a high level since, and in the last Five-Year Plan for the country (1981–1985) a total of M$1424 million or 18.6% of the total public expenditure on agriculture was spent on the facility. The fertilizer subsidy allocation for paddy between 1981 and 1985 was estimated at M$500 million. Briefly introduced in 1973–1974 when the world price of urea increased tremendously, it was reintroduced in 1979 and has since continued to benefit all paddy farmers who cultivate 2.43 hectares or less. Presently, the cash value of fertilizers provided to farmers through the scheme is estimated at M$232 per hectare per season. An equally generous price subsidy scheme said to benefit 162,000 paddy farmers throughout the country was estimated to cost M$175 million and M$186 million in 1984 and 1985.

The combined effect of state policies and support has raised yields and total output as well as brought about significant changes in the organization of paddy production. Peninsular Malaysia's total main-season wet paddy planted area increased from an annual average of 269,706 hectares in the period 1950–1955 to 369,546 hectares in 1974–1975. (See Table 9.2.) With improved drainage and irrigation, the off-season crop area increased considerably from an annual average of 2660 hectares to 199,945 hectares for the same period. Total output and yield also increased correspondingly. In 1950–1955, the average annual total output was 535,000 tons with an average yield of 1.87 tons per hectare. This later increased to about 1.6 million tons in 1971–1975 with an average yield of 2.8 tons per hectare.

The considerable changes that the paddy economy has experienced since 1960, however, have not been evenly distributed throughout the country and, in fact, vary widely between different rice-growing areas. It is possible to distinguish between two distinct geographical areas of paddy production which coincide broadly with the different manner in which the cultivation is carried out. The main and most productive rice-growing areas are located in the coastal plains in the northern states of Kedah, Kelantan, Penang, and Perlis and to a lesser extent in the central states of Perak and Selangor where climatic and soil conditions are more suitable for the cultivation than elsewhere in the peninsular. The seven major irrigated paddy schemes of West Malaysia are all located in this belt, and they presently account for more than half of the total planted area.[5] Since the cultivation of paddy here is geared largely towards the market, farmers tend to have larger farms, are more specialized in the crop, and derive a higher percentage of their household income from paddy earnings.

It is here too in the large irrigation areas that the development of the capitalist market in paddy has proceeded furthest, with much of the increased surplus devoted to production inputs such as fertilizer application, mechanization, and wage labor. The result of the development of productive forces in a situation where the unequal distribution of farm size and income has not been fundamentally addressed by state policies (Lim, Chapter 10) has been to widen inequalities and concentrate the benefits of the increased output amongst the larger landowning and operating class and a new class of intermediaries controlling inputs. These intermediaries include rural elites and bureaucrats who invested in farm machinery or used their official positions in public enterprises and farmers' organizations to obtain special privileges or perks for themselves. In particular, the investment in tractors and combine harvesters made possible by the implementation of the Green Revolution has facilitated the further accumulation of capital by the wealthy and has displaced labor previously supplied by poorer farm households in paddy transplanting and harvesting. Both these activities are performed almost exclusively by women who have been deprived of a major source of income.

In the process, tenant farmers have been the first group to be displaced in significant numbers as owners found it more profitable to cultivate than rent out land (Lim, Chapter 10). The outlook for small owner-operated farms that have traditionally been dependent on sale of their labor power for an important part of their income appears equally bleak. Initially benefiting from the yield increases, these households are now confronted with the prospect of competing with an emerging class of large capitalist farmers whose capacity to expand their operations has been strengthened by the rationalization of important parts of the labor process (Ikmal, Chapter 11). Thus, it is not surprising that we find the Gini coefficient of farmer income and wealth inequality highest and increasing in the Muda area, the most developed of the schemes (see Table 9.3).

Whilst the large irrigation areas exhibit one mode of development, the smaller inland valley areas located mainly in the central and southern states, but including the northwest region of Kedah and the interior of Kelantan, appear to be undergoing a different process of change. There, conditions for cultivation are generally less favorable, and as a result the paddy areas are small and interspread with other cash crops.[6] Since paddy is usually single-cropped, yields are lower than in the large irrigated scheme areas[7] and grown mainly for home consumption. In the central and southern regions the activity should be seen as an adjunct of a larger regional economy, which includes production of cash crops and participation in the wage active labor market located in the urban centers nearby. The more diversified economy of peasants in the inland valley regions is reflected by the higher contribution of off-farm income. A recent study of 2700 households found that about 50% of their total income was derived from off-farm employment (Malaysia 1981:39).

It has been observed that there has not been much change in the concentration of paddy land holdings in Negeri Sembilan, one of the areas typical of these smaller inland paddy regions; neither has there been an increase in tenancy or wage labor leading to widespread differentiation of the peasantry with respect to the ownership of paddy land (Kahn 1981:549). Instead, however, a process of general decapitalization or running down of the sources of capital found in this and other backward paddy areas seems to be taking place.

TABLE 9.3. *Distribution of Income and Wealth Among Farmers, Selected Paddy Double-Cropped Areas, Malaysia*

Area	Year of study	Form of income or wealth	Gini coefficient	Reference
Northwest				
Muda Scheme, Kedah-Perlis	1966	Farm size	0.35	Jegatheesan (1976:55)
Muda Scheme, Kedah-Perlis	1972–73	Farm size	0.36	Jegatheesan (1976:55)
Muda Scheme, Kedah-Perlis	1972–73	Gross annual income	0.41	Afifuddin (1978:347)
Muda Scheme, Kedah-Perlis	1972–73	Net farm income	0.41	Lai (1978:52)
Muda Scheme, Kedah-Perlis	1972–73	Net household income	0.38	Lai (1978:52)
Muda Scheme, Kedah-Perlis	1975	Farm size	0.38	Jegatheesan (1976:55)
Muda Scheme, Kedah-Perlis	1976	Farm size	0.40	Jegatheesan (1980:14)
Muda Scheme, Kedah-Perlis	1976	Farm size	0.44	Gibbons et al. (1981:157 fn)
Muda Scheme, Kedah-Perlis	1981	Farm income	0.43	Wong (1983:42)
West Central				
Tg. Karang Scheme, Selangor	1975–76	Net paddy income	0.25	Fredericks (1977:43)
Tg. Karang Scheme, Selangor	1975–76	Net paddy income	0.39	Fredericks (1977:43)
Tg. Karang Scheme, Selangor	1975–76	Net household income	0.28	Fredericks (1977:43)
Northeast				
Salor Scheme, Kelantan	1971–72	Farm size	0.24	Barker and Herdt (1978:10)
Lemal Scheme, Kelantan	1971–72	Farm size	0.27	Barker and Herdt (1978:10)
All Studies	n/a	n/a	0.34	n/a

Source: Taylor (1981:70, table 3.14).

There are several reasons which go to explain this phenomenon that manifests itself in the abandonment of paddy farms (by 1985 an estimated 106,000 hectares of paddy land throughout the country had been abandoned altogether) and out-migration from traditional villages. One is the inability of farmers working very small-size plots to reproduce themselves solely or mainly from paddy cultivation owing to the high costs of inputs and the low level of returns in areas other than the irrigation schemes. As rubber tapping, the other important agricultural activity, has suffered from low prices in the past few years, many farmers can be said to be experiencing a "simple reproduction squeeze." The southern and central farmers are also closer to the main areas of industrial and urban growth in Kuala Lumpur, Singapore, and the other smaller towns, and considerable numbers of the young and female members of farm households have been drawn into the new labor markets to provide cheap labor. Also many traditional villages in the land-rich state of Pahang are now surrounded by new non-paddy-land development schemes that offer relatively high and regular incomes. In these cases, out-migration has been so rapid that the remaining population is comprised largely of children and the aged (Salih et al. 1981). A third possible reason is the relative neglect by the state in the past which concentrated funds for paddy development in the large irrigation scheme areas with their considerable electorates. The decapitalization of land in these areas has lately become of considerable concern to the state. For this not only means degeneration of the surrounding economy but also massive out-migration (of youths particularly) into urban areas.[8] As in other Third World situations, the dependent industrial and service sectors in Malaysia have not grown fast enough to absorb this historically sudden inflow of rural labor. UMNO is particularly sensitive to this change for it is fearful of losing its rural clients to urban institutions that are not immediately under its control.[9]

During the past few years some effort has gone into developing new state-assisted agricultural programs for the region. The "reproduction squeeze" on small peasants and the attendant idle-land problem in marginal rice-growing areas, however, clearly offers new prospects for capital in agriculture. The state now mediates with owners of abandoned and idle land so that these small land lots can be rented out as large, contiguous parcels for rehabilitation by the private sector (including "graduate farmers," i.e., unemployed graduates who are encouraged to go into commercial farming). Of the 160,000 hectares of abandoned and idle paddy land in the country today, 124,000 hectares are to be rehabilitated by the private sector (62,000 hectares) and agencies under the Ministry of Agriculture (62,000 hectares). Nine private corporations are reported to have been involved in rehabilitation of abandoned and idle paddy land in the states of Kelantan, Malacca, Penang, and Perak (*Utusan Malaysia*, 2 December 1986). Thus, in contrast to the Muda area, large capitalist farms in these marginal rice-growing areas seem to appear only after petty producers have exhausted their resistance to market forces.

We had earlier noted that Malaysia is in many ways not a typical Southeast Asian country. Not only do the large size, relative affluence, and dynamism of its nonagricultural sectors stand out, but these sectors have also been responsible for generating a substantial flow of funds, part of which the state has transferred to the rural sector, and the paddy sector in particular. Despite this, substantial numbers in the rural sector still remain poor. In 1982–1983, the government estimated the number of rural poor households at 440,000 out of an estimated 800,000 rural households. The incidence of poverty amongst paddy households was estimated at 54% (Malaysia 1981:80).

Further, the scope for continuing improvements in the economic and social conditions of the peasantry in the immediate future does not appear as great as in the past. Within the paddy sector, rapid population growth, the small size of many holdings, and recent stagnation in yields, even in the largest irrigation scheme areas, could push many farmers deeper into poverty. At the same time, the difficult financial position the state presently finds itself in, despite large borrowings incurred locally and abroad, makes it unlikely that new policies will be as generous to the paddy community as they have been in the past. However, it is quite inconceivable that the government would risk losing a good deal of its electoral support amongst the rural Malays by stopping the fertilizer subsidy and paddy price support programs or by dismantling the elaborate and predominantly Malay-staffed bureaucracy providing services to the rural community. Clearly, if the present situation in Malaysia is not to deteriorate, a more searching analysis of the economic development strategy of both paddy sector and the country as a whole and stronger measures, including redistribution of productive assets, have to be contemplated rather than the intensified application of conventional development strategy.

Notes

1. For a fuller discussion of colonial policy and the development of plantations and peasant agriculture, see Lim (1977).

2. In the early 1960's the country had to import between 35% and 45% of its total supply of rice. Today, local production can meet up to 85% of national consumption.

3. The United Malays National Organization (UMNO) was the dominant political party in the multiracial Alliance coalition government at that time. It has maintained this position in the larger multiracial coalition known as the Barisan National formed in the early 1970's and which today is the ruling party.

4. The cost of producing rice in Malaysia is much higher than in the neighboring countries, especially Thailand. This explains why the huge amounts of money spent on drainage and irrigation and milling to increase peasant productivity have had to be supported further by price subsidies and import restrictions. Rampant smuggling of foreign rice into Malaysia is said to be resulting in a loss of about M$75 million in government revenue annually.

5. The seven major schemes are Muda, Kemubu, Sungai Muda, Krian, Sungai Manik, Seberang Perak, and Tanjong Karang.

6. In Malacca, for instance, one study found that paddy land formed only about 0.9 hectare or less than half the average farm area of 1.9 hectares compared to Kedah, Perlis, and Penang where paddy land was estimated to be about 72% of the average farm (Ho 1970:25).

7. A recent study found paddy fields of 1.51 and 2.24 tons per hectare in two small irrigation areas in Malacca, which is considerably below the average yield of 4.2 tons in the Muda area for the 1979 main and off season (Taylor 1981:97).

8. This explains why the average age of agricultural producers in Malaysia is estimated to be as high as 49 years (*Utusan Malaysia*, 29 January 1987).

9. UMNO's control in the countryside stems from its control over the District Office and the large number of government and quasi-government agricultural bodies that serve the rural population. See Scott (1985) and Shamsul (1986).

References

Ho, R. (1970). "Land Ownership and Economic Prospects of Malayan Peasants." *Modern Asian Studies*, vol. 4, 1, 83–92.

Kahn, J. (1981). "The Social Context of Technological Change in Four Malaysian Villages: A Problem for Economic Anthropology." *Man*, vol. 16, 4, 542–562.

Kementerian Pertanian (1985). *Perangkaan Padi Semenanjung Malaysia, 1983.* Kuala Lumpur.

Lim Teck Ghee (1977). *Peasants and Their Agricultural Economy in Colonial Malay.* Kuala Lumpur: Oxford University Press.

Malaysia (1976). *Third Malaysia Plan 1976–1980.* Kuala Lumpur: Government Press.

——— (1981) *Fourth Malaysia Plan 1981–1985.* Kuala Lumpur: Government Press.

——— (1984). *Mid Term Review of the Fourth Malaysian Plan, 1981–1985.* Kuala Lumpur: Government Press.

Salih, K., et al. (1981). *Laporan Pemulihan Kampong-kampong Traditional DARA.* Kuantan: DARA.

Scott, J. C. (1985). *Weapons of the Weak.* New Haven: Yale University Press.

Shamsul, A. B. (1986). *From British to Bumiputera Rule.* Singapore: Institute of Southeast Asian Studies.

Taylor, D. C. (1981). *The Economics of Malaysian Paddy Production and Irrigation.* Bangkok: Agricultural Development Council.

Chapter Ten

Reconstituting the Peasantry: Changes in Landholding Structure in the Muda Irrigation Scheme

LIM TECK GHEE

This paper examines changes in the structure of landholdings in the Muda area of northwest Malaysia, the largest and most productive rice-growing part of the country and the site of a M$245 million drainage irrigation project completed in 1974 with World Bank assistance.[1] Muda is also the region where commercialization of rice agriculture is most advanced and differentiation has gone furthest and is most complex.

The high degree of specialization in rice production coupled with the relative absence of diversification in the regional economy (see Lim and Ikmal, Chapter 9) implies that the question of access to and control over riceland is a crucial one determining the well-being of the rural population. Understanding the changes in landholding structure could provide important clues to our understanding of the processes of rural differentiation as they unfold in a relatively well-endowed capitalist economy.

The Historical Development of Rice Cultivation in Muda

Accounts of rice growing in Muda go back as early as the fifteenth century. Land was opened up by individuals or groups of peasants and it appears that by the early twentieth century when British colonial rule was established, there was already a sizable rice-growing community with well-established rights to the ownership and occupation of paddy land (Wilson 1958:8–9). Ordinary peasants owned as much as they could cultivate themselves with household labor, i.e., usually less than 2 hectares. But local and district chiefs and the state aristocracy as well as individuals with influence and money sometimes owned as much as 300 hectares each. These large tracts were operated with corvée, slave, or wage labor and/or rented out for payment in kind. According to Wilson (1958) the majority of cultivated or occupied land in Kedah carried "the general obligations of subjects to their Rajas comparable to the feudal system of Europe, with powers delegated to the territorial chiefs as Lords of the Manor. . . . Forced labor (*kerah*) was accepted as the lot of each cultivator, avoidable only by abandoning and thereby losing his land" (p. 8).

Control over strategic or productive tracts of land provided both the economic surplus and the sociopolitical influence that was essential for continuing domination by the Malay ruling elite. Although the small population in relation to the land during the precolonial period would have ensured fairly widespread distribution of ownership among free peasants, it is highly probable too that significant degrees of economic and social differentiation resulting from unequal ownership and control of agricultural land amongst individuals and groups were already well established prior to the coming of the British, given the political and economic importance of land.

New Land Opening

As the agricultural population increases and if the amount of land available for use remains static, peasants will be forced to eke out their living on decreasing plots of land. This could lead to a number of consequences, including attempts at maintaining standards of living through increasing self-exploitation, underconsumption by the peasant family, and enhanced conflict over access to land, resulting in displacement of the less powerful farmers.

In the Muda area, the population during the last 130 years has grown at a rapid pace that has begun to slow down only in the past 20 years. One source has estimated the population in the region in 1850 at 35,000 (FAO 1975:vol. I, 13). In 1980, the population was estimated at 336,400 or nearly ten times more, as a result both of a high natural rate of population growth and the influx of migrants from other parts of the peninsula, especially during the late nineteenth and early twentieth centuries.

Until quite recently the considerable impact that could be expected from the rapid population increase has been largely mitigated by the increase in cultivated area. It has been estimated that in 1851, approximately 10,530 hectares were in cultivation, giving a population/cultivated-land ratio of 1:0.3 hectares. Since then, land has been steadily opened up. For example, between 1931 and 1944 approximately 2025 hectares were opened up annually from a base of 64,300 hectares in 1931 to 89,100 hectares in 1940 (FAO 1975:vol. I, 13). It has only been in the last twenty-five years that the pace of new land opening has slackened, but even as late as 1980 the population/cultivated-land ratio in Muda stood at 1:0.29 hectares.

Whilst the overall population/cultivated-land ratio has changed little over the long historical time due to the relatively open land frontier,[2] there have been significant changes in the control over land exerted by various socioeconomic groups. These changes have been a result of several transmission and transaction mechanisms that operated in the past and are still operating today.

Transmission

There are two methods of division of inheritance amongst Malays: Islamic law and *adat* or customary practice. Both methods result in a spread of ownership as they provide that all the children of the deceased owner have a share in the land owned by the deceased. However, Islamic law distributes landed property unequally amongst male and female heirs, with males entitled to twice the amount received by females, resulting in land concentration along the male line. The *adat* method provides for equal shares.

Data from a comprehensive land tenure study in the Muda area in 1975–1976 (Lim et al. 1978; Gibbons et al. 1981) shows not only that male owners on average had larger holdings of paddy land compared with female owners (male average-sized holdings of 1.5 hectares, female average-sized holdings of 1.0 hectare) but also that there was a higher preponderence of male owners (56.5% of sampled owners were males compared with 43.5% female owners). A more recent village study (D. Wong 1983) suggests a trend towards use of the Islamic law of inheritance in preference to *adat*, but the impact of this on gender and production relations has yet to be studied.

The extent of dispersal of ownership brought about by transmission also depends on the number of beneficiaries and the time period between various rounds of transmission. A situation in which there are only single beneficiaries to landed property would result in the maintenance of the status quo as far as the number of owners is concerned. It could also be that only one heir inherits the property of both parents, resulting in a reduction of ownership dispersal, but these instances are relatively infrequent and multiple beneficiaries are more the norm in Malay society. Finally, the more time that elapses between several generations of farmers, the less dispersal of ownership there will be compared to when there is a rapid succession of generations.

Transaction

Sales transactions act as a countervailing force to the dispersal of ownership brought about by transmission. The volume of land transactions in Muda has fluctuated considerably according to the conditions prevailing in different periods, but the cumulative effect on the distribution of landownership has been considerable. The transfer of land from poor to rich peasants as well as from the peasant community to the wider outside community enhances the concentration of landownership through a decrease in the actual number of owners and an increase in variation in the area of land owned per landowner. It also brings about greater agrarian differentiation in the sense that former independent landowners are compelled to become either tenants or farm laborers at the same time as a class of large landowners emerges or becomes strengthened.

Trends in Concentration of Ownership in Muda: 1955–1975/76

Particularly in an undiversified economy like Muda, landownership is a vital determinant of agrarian structure, and it is of some importance to ascertain whether it has become more or less unequal over time and what the future trend is likely to be. The only statistical reference to the previous concentration of ownership in Muda comes from Wilson

(1958:67), who, in a much-quoted sentence, estimated that "not more than 2,000 families owned not less than two-thirds of the padi lands of North Malaya." The figure of 2000 families was a miscalculation on Wilson's part. He had arrived at the number from an analysis of a small sample of the estates of deceased landowners in which he found that 8.8% of the owners accounted for 67% of land. Using the figure of 248,920 holdings in North Malaya (that is, Kedah, Perlis, Province Wellesley, and Kelantan) obtained from one of his tables (p. 79), it can be calculated that in fact about 20,000 families and not 2000 owned 67% of the land, with the number in Muda alone amounting to 6445 families. Nonetheless, Wilson's assertion of a high degree of landownership concentration was well founded. According to our computation based on Wilson's data (tables 62 and 66), the Gini index of inequality was very high at 0.733 in 1955.

Apart from Wilson's finding on the high concentration of paddy landownership, a number of other scholars (notably Swift 1967:241–254 and Syed Husin Ali 1972:102) working on field studies in the 1960's had predicted that the ownership of paddy land would become even more concentrated over time. The argument behind this prediction emphasized increasing population pressure on a relatively fixed amount of land, stagnating agricultural technology, and the lack of alternative and additional employment opportunities for a large proportion of the paddy peasantry. These conditions were seen as forcing impoverished peasants to mortgage or sell their land to a few relatively wealthy people within and outside the peasant community, enabling the minority group to accumulate land rapidly. Thus greater concentration of ownership of paddy land was seen as inevitable.

An exhaustive study of land tenure in Muda twenty years after Wilson's pioneering work, however, produced some surprising results.[3] First of all, the pattern of distribution of paddy landownership in 1975–1976 was found to reveal a wide dispersal of ownership (Table 10.1). Smallholdings (below 1.15 hectares) comprised 61.8% of all holdings and 26.5% of area, while at the other extreme large holdings (2.88 hectares and more) accounted for 11.2% of all holdings and 42.3% of area. Even though the distribution of landownership was unequal (Gini index of 0.538), it was not as extreme as that estimated by Wilson (1958:63–67) for 1953. When compared with Wilson's findings, this evidence suggests that concentration of ownership of paddy land in the area probably *decreased* considerably over the period 1955 to 1975.

TABLE 10.1. *Size Distribution of Paddy Landownership:*[a] *Muda Irrigation Scheme Area, 1975–1976*[b]

Size class (hectare)	Number of holdings (%)	Relative frequency (%)	Area of holding (hectare)	Relative frequency (%)
0.01–0.57	17,735	39.3	4,735	7.8
0.58–1.15	10,163	22.5	8,303	18.7
1.16–1.72	6,556	14.6	9,252	15.3
1.73–2.30	3,278	7.3	6,511	10.8
2.31–2.87	2,363	5.2	6,112	10.1
2.88–11.50	4,835	10.8	22,126	36.6
11.51 and above	185	0.4	3,442	5.7
Total	45,115	100.1	60,481	100.0

Source: Gibbons et al. (1981:22).

[a] Data were collected from the Land Offices and aggregated through the I.C. number of the owner. They excluded 4135 hectares of paddy land owned by the state, 949 hectares owned by banks and private companies, and 39,436 hectares owned by individuals for whom no I.C. number could be obtained. Strictly speaking, therefore, the data cover only 57.6% of the total paddy land in the Muda Irrigation Scheme Area. However, a comparison of the central tendency and dispersion of the size distributions of included and excluded shares showed considerable similarity in the two distributions. The mean size of included shares was 0.81 hectare compared to 1.02 hectares for those excluded. The standard deviations of the two distributions were 0.89 hectare and 1.13 hectares respectively. Given the basic similarity of the two distributions and the large absolute number of shares included, there seems little reason to doubt the generality of these findings on the size distribution of paddy landholdings for the whole Muda Scheme Area. Based on these findings, it is estimated that there are about 78,000 discrete paddy landholdings in the Muda Region. To estimate the number in any size category, multiply its absolute frequency by 1.7289.

[b] Gini index = 0.538; mean = 1.34 hectares; standard deviation = 2.22 hectares; median = 0.78 hectare.

Why did early predictions of increasing concentration of ownership of paddy land fail to materialize, at least until 1975–1976? These patterns are particularly interesting in view of widely held notions about the adverse impact of the Green Revolution on landownership distribution.

Long-term fluctuations in the intensity of the forces of concentration constitute part of the explanation. In 1955 when Wilson was doing his study, the agricultural economy of Muda, although not yet highly developed, had undergone commercialization over a long period of time. As early as the nineteenth century there were reports of rice exports from Kedah to Penang and other neighboring states. There was also a generally low level of development of the productive forces, and the frequent failure of crops due to water shortage or pest infestation afforded much opportunity to the richer peasants and trader-financiers to dispossess the poorer peasants of their land. Reports by colonial officials during the late nineteenth and early twentieth centuries indicate that indebtedness was very prevalent. There was widespread practice of *padi kuncha*, a system of promise of sale of a part of the paddy crop at lower prices than the prevailing market price, and many variations in credit procurement using the land or crop as security, with default on loans frequently resulting in loss of the land by the owner to the creditor.

The processes of land dispossession and accumulation resulting in the concentration of ownership would have continued unabated until the 1930's when an important check was introduced by the colonial government. This check was in the form of restrictive legislation which prohibited land transfers from Malays to non-Malays as well as alienation of state land within a gazetted reservation area to non-Malays. It was clear to the British that exploitation of the peninsula's resources could be carried out without the need to interfere harshly with Malay economic interests, especially in paddy. There was a plentiful supply of virgin land, and cheap labor for non-rice-farming activities was available through the importation of Chinese and Indian immigrants. Hence, it seemed eminently sensible to delineate separate geographical enclaves for the Malays where they could pursue their economy and society relatively unhindered.

Quiescent or passive Malay communities in reservations were an important insurance for British political domination, and they also had an economic bonus in that these communities would be producing rice, the staple food of the country, and could be prevented from competing with non-Malay—primarily British—rubber growers. In retrospect, the reservation legislation was a stroke of political genius on the part of the British, securing their economic and political objectives vis-à-vis the Malays and balancing Malay and non-Malay interests by diverting the latter into areas where they would not largely conflict with the indigenous people.

The immediate and long-term effect of the legislation (which is also regarded by the postcolonial Malay-dominated government as an important bulwark protecting the Malay community against the economically more assertive non-Malays) in Muda and other Malay reservation areas in the country was effectively to prevent Malay lands from being accumulated by non-Malays. But it did not prevent land accumulation by the richer Malays to whom foreclosure of land belonging to poor

Malays provided much opportunity to accumulate land cheaply in a restricted market. However, up to the 1950's, when new land development continued through the extension of drainage and irrigation works, there was access to new riceland for disposessed or small peasants with sufficient labor and capital resources to clear the forests. Thus, the process of land accumulation by the rural Malay elite did not produce a large class of landless Malays although it did produce a distribution of ownership that was considerably skewed.

Since Wilson's time, especially after the successful implementation of double-cropping, the possibilities for dispossession and accumulation of land in the Muda area have diminished. Production increases have expanded substantially the income available to all farming households, including poor ones; more important, the stability in production, increased frequency of income in the 1970's, and access to alternative sources of credit have resulted in the virtual elimination of the *padi kuncha* system and reduced the number of usufructuary loans and mortgages (*gadai* and *jual janji*), thereby making it less easy for poor peasants to be dispossessed of their land. Also the price of paddy land has moved up sharply and has made it difficult for buyers of land to accumulate large holdings through purchase.[4] Finally, inheritance would have acted in many cases to disperse medium and large concentrations of ownership.

The existence of these checks on the forces of land dispossession and accumulation which operated so strongly in the previous period do not imply that peasants are not being dispossesed. Some poor peasants continue to sell off their land to raise cash for unexpected or unusual needs, but it is unlikely that the degree of concentration of ownership will increase sharply in the foreseeable future, unless new forces come into play. While concentration of ownership leading to the dispossession of poor landed peasants does not appear likely in the immediate future, the dispersal of ownership amongst an even larger number of small owners leading to tiny holdings is a distinct possibility.

At present 60% of all holdings are below 1.15 hectares or what has been generally regarded as the minimum farm size necessary to sustain a peasant family above the Malaysian poverty line, assuming it owns and operates all of this area and is entirely dependent upon rice income for its livelihood.[5] Over the next decade, assuming a continuation of Malay inheritance practices, there must be a substantial movement of small owners off the land or the implementation of measures that can effectively bring about a more equal distribution of land owned if hold-

TABLE 10.2. *Size Distribution of Operated Paddy Farms: Muda Irrigation Scheme Area, 1955 and 1975–1976*

Size class (hectare)	Percentage of farms		Percentage of area	
	1955 (N = 46,547)	*1975–1976 (N = 61,164)*	*1955 (N = 95,950 ha)*	*1975–1976 (N = 99,002 ha)*
0.01–0.57	13.6	20.7	2.2	4.2
0.58–1.15	18.8	25.9	7.9	12.8
1.16–1.72	20.3	19.5	14.0	16.7
1.73–2.30	15.0	11.1	14.5	13.4
2.31–2.87	14.4	8.3	18.4	13.0
2.88 and above	17.9	14.5	43.0	39.9
Totals	100.0	100.0	100.0	100.0

Sources: Berwick (1956: table 6); Gibbons et al. (1981: table 48).
Note: The data here refer to the "Main Kedah Plain," which corresponds roughly to the Muda Scheme Area, except that part of Perlis which is in the Scheme Area is excluded (about 15,525 hectares) whilst a part of the district of Kuala Muda not in the scheme area is included (about 6469 hectares).
It should also be pointed out that the farm size categories here are 0.029 hectare greater than in the 1975–1976 categories. Given the tendency of farmers to round-off their farm size to the nearest *relong*, the local measure, this might have inflated somewhat the small and medium farm size categories for 1955. Gini index = 0.396.

ing sizes are not to decline further and the number of very small owner-farmers is not to increase more sharply.

Distribution of Farms and Farmers in Muda: 1955–1975/76

The distribution of operated paddy land (farms) and farmers is an equally important question since it is access to land use rather than ownership per se that largely determines the economic well-being of the agricultural population. Here we have some trend data from two large-scale studies conducted in 1955 and 1975–1976 (Berwick 1956 and Gibbons et al. 1981). The geographical areas covered by the two studies are broadly the same so that it is possible to compare their data on farmers and farming area directly. From Table 10.2 it can be seen that in 1955 there were an estimated 46,547 farms in Muda with a mean farm size of 2.06 hectares. Small farms, which we have defined as below

1.15 hectares, comprised about one-third (32.5%) of all farms and 10.1% of area compared with medium-sized farms (1.15–2.85 hectares) which comprised about half (49.6%) of all farms and almost the same proportion (46.9%) of area.

Since then, there has been an increase in the number of farms to an estimated 61,164 in 1975–1976 (Table 10.2) as a result of the vigorous population growth, only partially alleviated by out-migration. Small farms were most numerous (46.7%) followed by medium-sized farms (38.9%) and large farms (14.7%). As with paddy landownership, distribution of farms amongst the various size categories was unequal with the Gini index value at 0.445, indicating a slight increase in the concentration in paddy land farmed or operated (as distinct from paddy land owned) during the twenty-year period. Small farms occupied 17% of the operated land compared with 43.2% for medium-sized farms and 39.8% for large farms. The moderately high degree of concentration of operated land can also be seen from the fact that whilst the overall average farm size was 1.61 hectares, that of small farms was 0.57 hectare compared with the 4.4 hectares average size of large farms. The median farm size was only 1.18 hectares.

When comparing the 1955 and 1975–1976 data on farm size, one of the most striking features is the increase in the number of small farms from 15,000 to more than 22,800. The reasons derive mainly from the successful distributive impact of the Green Revolution.[6] Although double-cropping has obviously generated greater absolute benefits for large operators, it has also resulted in real production increases for small farms. Apart from irrigation and high-yielding-variety seed strains, small farmers have also been able (though less easily than large farmers) to avail themselves of fertilizers, subsidies, credit, and other institutional support from the government through a number of governmental and quasi-governmental agencies set up to provide low-interest credit, assist in marketing, eliminate milling abuses, and provide various other services to the paddy community. Of these, the most important is the National Padi and Rice Authority (Lembaga Padi dan Beras Negara or LPN), which has an extensive network of rice mills and controls rice prices through a variety of measures, including buffer stockpiling, licensing of import quotas, and control of sale by millers and wholesalers.

Along with the gains in small-farm productivity, the increase in the demand for labor during the early years of the Green Revolution benefited poor households with inadequate land but surplus labor. In 1975–

1976, transplanting was still entirely a manual operation and about 75% of the harvesting and threshing work depended on human labor. The extent of supplementary income derived by the sale of labor power by small farmers is suggested by a 1972–1973 study which found that 24.3% and 20.0% of the mean incomes of households earning less than M$600–M$1200 respectively were derived from off-farm sources, mainly agricultural labor wages (Lai 1978:39).

Some of the gains in income to farmers also occurred as a result of the increase in paddy prices from M$13.12 per pikul (approximately 60.5 kg) in 1966 to M$26.66 in 1975. According to one estimate, the improved prices (which were brought about mainly by the spiraling international rice prices in 1973) accounted for nearly half (47%) of net income increase in Muda between 1966 and 1975 and 93% between 1972–1973 and 1975 (Jegatheesan 1977:64). Nevertheless, the positive role of the Green Revolution technology in enabling small farmers to survive must be acknowledged, especially since there was hardly any new land opening in the 1970's to accommodate the growing population. According to census data, Muda's population grew at an estimated 1.54% between 1957 and 1970 and 1.58% between 1970 and 1980, a figure about half the national rate due to the substantial amount of out-migration from Muda. Without the gains in productivity generated by the Green Revolution, small farms would have existed under worse conditions or might not have been able to avoid being absorbed by the larger and more well-to-do farms.

Apart from the growing preponderance of small farms, technological change and state subsidies have also been associated with a sharp decline in tenancy. Whereas owner-operators and, to a lesser extent, owner-tenants increased their share of farm numbers and area since 1955 (the former both proportionally and absolutely, the latter only in absolute terms), tenant farms declined by 24% in number and 41% in area. The reduction in numbers and area was most marked among medium and large tenants.

The decline in tenancy is highly significant in view of the kin-based bilateral settlement system predominant in Muda and kinship links that have long protected tenancy arrangements. The 1975–1976 study found that 71% of non-owner operated parcels and 63% of paddy land in the Muda area were operated by kinfolk of the owners. Although kin relationships may have ensured continued access to land for some operators, the pressures exerted by population growth and the increasing profitability of owner operation due to production and paddy price increases have made tenants in Muda especially vulnerable to displacement.

One possible way in which displacement could have come about is through death of owners of land operated by medium and large tenants and the subsequent inheritance and operation of this land by the beneficiaries who, because of Malay inheritance practices, would have numbered more than the deceased owners. Thus the size of the new owner-operated farms would be smaller. Another way would be for owners of land tenanted by medium and large farmers to take back the land to operate part of it themselves and to let a close relative (perhaps the former tenant) operate the other part rent-free, thus creating two small farms, one owner-operated and the other under familial tenancy. Some medium and large owner-operators may have taken back land from tenants, thus increasing the farm sizes of the former. Finally, some of the percentage drop in medium and large tenants may indicate only a relative decline in their position. The total number of paddy farmers in the Muda area probably increased by about 25% between 1955 and 1975–1976 (from about 46,500 to 61,200), and average farm size declined from 2.0 to 1.6 hectares over this period. If most of the new farmers were small owner-operators, perhaps on land they had cleared themselves, and if none or very few were medium and large tenants, then the relative position of the latter, rather than their absolute numbers, would have declined while that of small owner-operators would have increased.

Small owner-operators have indeed become by far the largest single group of paddy farmers (33.8%). Together with small tenants and small owner-tenants who accounted for 16.3% of all paddy farmers, the three categories of small farmers (owners, tenants, and owner-tenants) add up to more than half (50.1%) of all paddy farmers although operating only 19.5% of all the paddy land. With an average farm size of 0.6 hectare, these small farmers would have had an income below the Malaysian poverty line if they had been primarily dependent on paddy production.

The problems of access to paddy land experienced by small farmers whatever their tenure are to a considerable extent associated with the continuing concentration of access to paddy land by medium and large farmers. Although the proportion of area operated by larger farmers declined in the twenty-year interval, it is still quite high. The maintenance of disparities in the operation of land is suggested by the rise in the Gini index of operated holdings in the Muda area from 0.396 in 1955 to 0.445 in 1975–1976.

Leaving aside the tenants displaced entirely from farming in Muda, whose fate we do not know, the farmers that were on the land in 1975–1976 were generally in a better position than their counterparts in 1955 as a result of double-cropping that virtually doubled their production and government assistance in paddy pricing, provision of fertilizers, milling services, etc. However, large numbers still lived off small-sized farms below the country's poverty line. The government's failure to come to grips with the central issue of concentration of ownership and operation meant that there was no dramatic progress in its stated goals of reducing inequalities in the countryside and eradicating poverty.

Trends in the Organization of Production Since 1975–1976

No data are available on trends in landownership or operation since 1975–1976, but recent evidence on the organization of production is suggestive of the evolving pattern of inequality. The data come from two studies conducted in 1981, one dealing with large farms (Lim Teck Ghee 1984) and the other, which focused on the status of the farm household economy, providing information on small farms (Wong Hin Soon 1983).[7]

These data show first that large farms average only 60% of the total labor (on a per unit land basis) used by small farms. This decline in total labor use with increasing farm size reflects the rapidity of the spread of increased mechanization in the larger farms in contrast to the more intensive use of labor, especially family labor, in smaller farms. Mechanization has made inroads into farming practice both on small as well as large farms. Small farms generally depended on the use of hired machinery to supplement labor inputs in plowing and harvesting, with only a small proportion reporting machine use for nursery work, leveling, and transportation. In contrast, large farmers employed the use of both their own as well as hired machines in four of the five stages requiring mechanical assistance, i.e., nursery work, plowing, harvesting, and transporting. Own machinery (mainly in the form of four-wheel tractors) was more employed in land preparation than hired machinery, but there was a high dependence on hired machines (combine harvesters and lorries) for harvesting and transportation.

The spread of mechanization since the mid-1970's has been especially spectacular in harvesting, which was once the largest user of labor but has now become almost completely mechanized. Wong's study shows that only 10.5% of farmers continue to harvest their paddy manually, and this covered only 7.1% of area (1983:8). The implications of this for the wage-labor income of small peasants are well spelled out by Scott in his recent work based on the Muda area. According to his calculations in 1980, "based on the share of cutting (usually women's work) and threshing (usually men's work) in total hired labor and the intensity of combine-harvester use . . . combines have cut paddy wage labor receipts by 44 percent. For the poorest class of small farmers, this represents a 15 percent loss of net income in the case of tenants and an 11 percent loss in the case of owner-operators" (Scott 1985:75–76).

The remaining source of high labor demand in paddy in the Muda area is transplanting when seedlings have to be pulled, transported to the transplanting site, transplanted, and replaced with new seedlings when they do not take root. On both small and large farms, own and hired labor in transplanting totaled more than 50% of the time expended in all activities connected with the cultivation.

More than 20% of the labor of small farms and 85% of large farms came from hired labor used mainly in the transplanting process. But it does not appear that small farms are the main suppliers of hired labor. Wong in his 1981 study has shown that a small number of farmers undertake work on other farms for cash (27% of those who hire in labor hire out their own or their family labor for farmwork) and has concluded that this "can only indicate an unwillingness of the farmer and his family to enter the labor market" (p. 21). Given the poverty of many small farmers—in Wong's sample, 55.6% of all small farmers were found to earn paddy incomes that put them below the poverty level (p. 29)—this does not seem plausible. A more reasonable explanation for the apparent unwillingness of small farmers to enter the labor market is the very short time frame of transplanting and other farm activities in Muda due to the infrastructural development which makes it difficult even for small farmers to satisfy the labor requirements on their own farms. Thus, whilst small farmers use almost ten times the amount of family labor per hectare compared with large farms, they have had to use about 40% of the hired labor used by large farms, in addition to their considerable use of machines.

Wong's data suggest that most of the hired labor came from local sources (92%) with much smaller amounts coming from Thailand/Kelantan (7%) and areas fringing the Muda scheme (less than 1%). Since the local sources are not mainly from the small farms, it is possible that they are the displaced tenants of the Green Revolution who are now in the nonagricultural informal sector and return to the fields during periods of peak labor requirements to earn extra income. More research needs to be carried out on this group, especially since the loss of this wage source will further adversely affect their economic position should broadcasting techniques and mechanical harvesters become widespread.

Meanwhile, the yields of small and large farms, including four different tenurial groups, have converged. In 1972–1973, one study had found that productivity per hectare varied inversely with farm size with reported average yields of large farms (5 hectares and over) at only 56% of the smallest (below 1 hectare) and 69% of farms between 1 and 2 hectares (Lai Kok Chew 1978). By 1981, the yield gap had narrowed considerably with small farms as a group producing only 1% more paddy per hectare. While the narrowed yield gap indicates the more considerable gains the large farms have reaped from the Green Revolution, it is equally important to note that the development of large commercialized farms has not yet resulted in a yield gap in favor of the large farms.

That this has not happened is probably due to a number of factors. Our labor data showed that small farms put in considerably more labor to work their land. This would produce a positive impact on yields. Smaller farmers also seem to have been able to secure as much access to subsidized fertilizers, a crucial production input, thus enabling them to reach fairly high levels of yields and preventing any gap in yields opening up between them and large farmers.[8] Small farmers also have easy access to machinery in the form of hired tractors and combine harvesters. Whether the introduction of further yield-increasing technology involving large-scale economies in the future will bring about a divergence in yields between small and large farmers remains to be seen. Of more immediate concern is the finding that yields for all farmers appear to have reached a ceiling. Yield figures for Muda over the past eight years in fact show a stagnation,[9] but whether this is due more to technical reasons or to declining availability of willing labor is not clear. Wong's study for small farms, which shows that the per capita time spent by household members involved in agriculture is only 117 man-hours (which works out to 14.6 full working days in a season, a very low figure by international standards), implies the latter reason.

Conclusions

The first stage of the Green Revolution in Muda resulted in the reduction in the number of tenants. Production and productivity gains were to no avail to this class of operators when owners of land demanded their land back to operate themselves or with kin labor. The tenants that remained appear to belong to two quite different groups. One group consists of tenants with kinship ties to landlords, and they have the partial protection afforded by these powerful bonds. The other consists of highly entrepreneurial farmers who have been able to bid for leases (often long ones) at comparatively high rents. Should profitability of the cultivation increase or if population pressure on the land worsens—as appears more likely—tenancy should decline even further.

It does not appear that most displaced poor tenants have become full-time farmworkers within Muda. Mechanization has reduced labor demand at the same time as labor needs have been collapsed into a few months a year as a result of the new varieties and drainage schedule. Hence, it is more likely that displaced tenants migrated permanently or now sell their labor to a variety of places, including the paddy sector during periods of peak labor demand.

We have also seen how capitalist farmers (characterized by operation of large units of land, use of hired labor, and ownership of farm machinery) have been able to push back the management diseconomies usually associated with rice cultivation through changes in cultivation techniques. However, whether they can expand at a rapid rate or not will depend on a number of factors including the relative profitability of rice production against other forms of investment and the extent of marginalization of small farmers that might afford opportunities for accumulation of leases. Although the technical conditions for the expansion of large-scale capitalist farming in Muda will become more favorable in the near future as a result of further public infrastructural investment in some parts that will increase canal, drain, and farm road density from the current 10 meters per hectare to 30–35 meters per hectare, it is highly unlikely that the social conditions will be as conducive. This is partly because there are also special obstacles in the path of Chinese capitalist farmers, generally acknowledged to be the most entrepreneurial. The present generation of Chinese farmers appears to be the children of merchants and rice millers who have gone into rice production as much through circumstance (default by debtors, etc.) as by choice. Although they have accumulated much within the limits imposed by constraints such as the Malay Reservations Enactment, it is unlikely that the government, for political and social reasons, will permit them to expand their operations to any considerable extent. Malay capitalist farmers should find it easier to expand their farms compared with their Chinese counterparts but, given that the small Malay peasantry is the mainstay of the ruling Malay party's electoral support, they will probably be encouraged to diversify outside agriculture so as not to place pressure on the small farmers. In this, they will be strongly supported by state policies that have emphasized the development of a Malay commercial and business class.

Owner-operators initially benefited through the resumption of tenanted land and, especially for labor-surplus households, the heightened labor demand from large farms. Since then, labor demand has fallen considerably but the losses from this source have been offset by increasingly expensive subsidies made available to the paddy sector. The modified fertilizer subsidy program, for example, provides operators with fertilizers equivalent in value to M$231.58 per hectare, up to a maximum of 2.43 hectares, although larger operators have no difficulty claiming for their entire farm (Lim 1984:18). Similarly, the price support subsidy was sharply increased from M$2 per 60.5 kg of paddy in 1980 to M$10 whilst the minimum support price was raised from between M$24–M$28 for 60.5 kg of short, medium, and long grain in 1974 to between M$26 and M$30 in 1980. (See also Lim and Ikmal, Chapter 9.)

These subsidies have been an important reason explaining not only the survival of small farms but also perhaps their proliferation. Based on Wong Hin Soon's (1983) figures (tables 17 and 30, pp. 27 and 36 respectively) the cash subsidy alone accounted for 37% of total net paddy income and 19% of total income from all sources during the 1981 off season. At the same time it is obvious that the subsidies have benefited the largest farms more since the amounts given out under the various programs in effect are directly related to farm size. In the process, income distribution amongst Muda's farmers has worsened, and the gulf between large capitalist farmers and small peasants has widened even more.

Despite the subsidies, many of Muda's small farmers still live on the margins of poverty as a result of their small farm size and the lack of alternative employment opportunities. The refusal of government to effect any radical changes to the inequities in landownership and farm size implies that their well-being will, to a large extent, be dependent on the structural changes taking place in the other sectors of the Malaysian economy. If the primary sector can rapidly contract relative to the secondary and tertiary sectors, and the workforce in agriculture can cease to grow altogether, then it could be that the agrarian problem in Muda and elsewhere in the country can be contained. In the late 1970's this seemed achievable when the country's economy was buoyant. However, there are ominous signs now that the larger economy is in considerable difficulty as seen in the sharp decline in the prices of the major commodity exports, the rapid increase in public debt, and the slowdown in manufacturing and industry. Barring an extraordinary recovery of the Malaysian economy in the remaining half of the 1980's, the scenario painted by one optimistic observer of the Malaysian agrarian problem within one generation becoming more like that "confronting France and Italy over the past decade or more: how to give decent income to a declining peasantry through a system of subsidies paid for by taxing other, presumably richer, social groups" (Bell 1980:210) seems quite unrealistic.

Notes

1. The scheme with an irrigated area of almost 100,000 hectares during the main season is more than four times as large as the country's next largest scheme. Since the completion of Muda I, a further stage of infrastructural development aimed at benefiting a quarter of the total farming area was initiated. The second-stage project is estimated to cost M$230 million, not including the annual operation and maintenance costs, which are considerable. In 1975–1976, the operation and maintenance costs in Muda were estimated (at M$19 per hectare) at M$1.8 million (Taylor 1981:129).

2. It should be noted that the newer settlers have had to colonize less fertile and more inaccessible lands. About 25% of the existing area suffers from salinity and sulfurous conditions.

3. The study consisted of a survey of all landownership data available from land offices as well as an exhaustive field survey of all farms in the Muda scheme area. See Lim et al. (1978) and Gibbons et al. (1981).

4. In 1975 if one assumed a price of M$5000 for a *relong* (the local measure equivalent to 0.288 hectare) of average-yielding paddy land, a buyer would need M$50,000 to accumulate 10 *relong*. This was a sum well beyond the reach of all but the richest peasant.

5. This farm size poverty line has been calculated from Jegatheesan (1977:30, table 2.7), who shows that the average annual net agricultural income per *relong* in 1975 was M$619.14. Since the minimum income required to sustain an average-size family above the poverty line was officially set at M$2400 per annum, a farm of at least 4 *relong* would be needed to ensure that a peasant family live above the poverty level.

6. A succinct analysis of the distribution impact of the Muda scheme amongst different farm groups is provided by Lai Kok Chew (1978:38–57).

7. In these studies, large farms were defined as those above 5.75 hectares whilst small farms were defined as those below 1.7 hectares.

8. Whilst small farmers generally do not appear to have been discriminated against in the Fertilizer Subsidy Program, there have been several reports of generalized discrimination against members of the main Malay opposition party.

9. From 1971 to 1981, dry-season yield fluctuated more than wet-season yield due to drought conditions during several years.

References

Akimi Fujimoto (1983). *Income Sharing Among Malay Peasants: A Study of Land Tenure and Rice Production*. Singapore: Singapore University Press.

Bell, Clive (1980). "The Future of Rice Monocultures in Malaysia." In D. C. Reining and B. Lombard, *Village Viability in Contemporary Society*. Boulder, Colorado: U.S. Department of Agriculture.

Berwick, E.J.H. (1956). *Census of Padi Planters In Kedah, 1955*. Alor Setar: Department of Agriculture.

de Koninck, Rudolphe (1985). "Is Land Ownership a Burden." *Kajian Malaysia* (Penang), vol. III, no. 1 (June), pp. 136–154.

Federation of Malaya (1956). *Final Report of the Rice Committee*. Kuala Lumpur: Government Printing Office.

Food and Agriculture Organization–World Bank (1975). *The Muda River Study: A First Report*. 2 vols. Rome.

Gibbons, D. S., et al. (1981). *Land Tenure in the Muda Irrigation Area: Final Report, Part II: Findings and Implications*. Penang: Center for Policy Research.

Griffin, K. (1974). *The Political Economy of Agrarian Change: An Essay on the Green Revolution*. London: Macmillan.
Jegatheesan, S. (1977). *The Green Revolution and the MUDA Irrigation Scheme*. Monograph no. 30. Telok Chengai: MADA.
Lai Kok Chew (1978). "Income Distribution Among Farm Households in the Muda Irrigation Scheme: A Developmental Perspective." *Kajian Ekonomi Malaysia*, vol. 15(1), June, Kuala Lumpur, pp. 38–57.
Lim Teck Ghee (1984). "Small and Large Padi Farms in Muda: A Comparison of Organization of Production and Yield." Unpublished paper. Penang: Center for Policy Research, Universiti Sains Malaysia.
Lim Teck Ghee et al. (1978). *Land Tenure in the Muda Irrigation Area: Final Report, Part I: Methodology*. Penang: Center for Policy Research.
——— (1980). "Accumulation of Padi land in the Muda Region: Some Findings and Thoughts on Their Implications for the Peasantry and Development." Unpublished paper presented at the conference on The Peasantry and Development in the ASEAN Region, Universiti Kebangsaan Malaysia, Bangi.
Muhammad Ikmal Said (1985). *The Evolution of Large Paddy Farms in the Muda Area, Kedah*. Penang: Center for Policy Research, Universiti Sains Malaysia.
Scott, J. C. (1985). *Weapons of the Weak*. New Haven: Yale University Press.
Swift, M. G. (1967). "Economic Concentration and Malay Peasant Society." In M. Freedman (ed.) *Social and Economic Organization: Essays Presented to Raymond Firth*. London: Frank Cass.
Syed Husin Ali (1972). "Land Concentration and Poverty Among the Rural Malays." *Nusantara* 1 (Jan.), Kuala Lumpur, pp. 100–113.
Taylor, D. C. (1981). "The Economics of Malaysian Paddy Production and Irrigation." Bangkok, Agricultural Development Council.
Wilson, T. B. (1955). "The Inheritance and Fragmentation of Malay Padi Lands in Krian, Perak." *Malayan Agricultural Journal*, vol. 33, no. 2, pp. 78–91.
——— (1958). *The Economics of Padi Production in North Malaya*. Bulletin no. 103. Kuala Lumpur: Ministry of Agriculture, Federation of Malaya.
Wong, Diana (1983). "The Social Organization of Peasant Reproduction: A Village in Kedah," Ph.D. thesis, University of Bielefeld.
Wong, Hin Soon (1983). *MADA II Evaluation Study—An Impact Evaluation Study of the Muda II Irrigation Project*. Telok Chengai: MADA.

Chapter Eleven

Large-Farmer Strategies in an Undiversified Economy

MUHAMMAD IKMAL SAID

The most salient feature of rice-growing communities in Southeast Asia is the predominance of small farms. This is true not only in areas where rice cultivation coexists with a range of other activities, but also in more specialized and less diversified regions like the Muda Plain in northwest Malaysia.

Efforts to characterize such economies have typically focused on the numerical importance of small family farms. Kahn (1980, 1982), for instance, insists that many rural Southeast Asian communities are best understood in terms of petty commodity production. Bray (1983), who invokes the case of Muda, maintains that wet rice production is inevitably associated with small-scale family farming because of its peculiar technological requirements.

In these and other analyses, the role of large farmers has been largely ignored. Although a majority of farmers in the Muda area operate extremely small farms (Lim, Chapter 10), a few large-scale producers do exist. The influence of these farmers extends well beyond their number and the proportion of land area under their control, and theories that ignore their presence are bound to be misleading. This paper describes the strategies of large-scale producers in the Muda Plain and considers some of the broader implications of these patterns.

Large Farms in the Muda Area

In 1975–1976, farms below 1.15 hectares formed 45.7 percent of the total number of farms but only 17.3 percent of the operated area in Muda. On the other hand, their wealthier counterparts who operated more than 5.75 hectares formed only 2.4 percent of farms but 11.3 percent of operated area (Gibbons et al. 1981:98). Farms above 10 hectares are much fewer, of course. My recent fieldwork (completed in late 1985) on the largest known farms in the Muda area located forty that were above 10 hectares (five farmers refused to be interviewed). Thirteen of these farms were above 15 hectares. Informal discussions with these wealthy farmers suggest that the actual number of farms above 10 hectares may be twice the number located. Thus, these large farms probably constitute about 0.13 percent of the total number of farms (61,000). Although these large farms are an exception, their operators exercise considerable influence in the rural economy.

Data in the following discussion are drawn from two studies; the first was carried out in 1981 (Lim 1985; Muhammad Ikmal Said 1985a), while the second I completed in late 1985. In the 1981 study, large farms were defined as those that were more than 7.19 hectares. I selected operators of the largest known (and willing) Chinese and Malay farms in the Muda area in the 1985 study. Twenty-five operators of large farms from each community were selected from four sources: the 1975–1976 survey of landholdings by Gibbons et al. (1981), the large-farm study of 1981 (Muhammad Ikmal Said 1985a), the National Paddy and Rice Authority (or Lembaga Padi dan Beras Negara, LPN for short) registration list, and finally the LPN price-subsidy list. A total of 125 farms from these four lists were recorded to be above 8.63 hectares. The largest farms were operated by Chinese farmers. As the range between the largest Malay and Chinese farms was rather large, I decided to select only farmers operating more than 10 hectares for in-depth interviews. Twenty-four Chinese farms and nine Malay operators had farms above 10 hectares.

Contrary to popular belief, large farms are not a new phenomenon. They are known to have existed during the colonial period. Berwick's census of rice cultivators in the state of Kedah in 1955 (1956:20–23, 30–32) shows that farms above 4.3 hectares (which constituted about 5 percent of total farms) constituted a total of 20,432 hectares or 21.3 percent of the total area. This gives a mean farm size of 8.5 hectares for those farms above 4.3 hectares. This is larger than the corresponding mean farm size for 1966 (5.78 hectares) and 1975–1976 (6.13 hectares). The agricultural census of 1960 (Malaya 1961) shows more clearly the existence of such large rice farms; there were sixty-four farms (or 0.14 percent of the total number in Kedah) that were above 10.1 hectares.

It is not clear how these large farms emerged from and coped with the trying conditions of the fifties, when the rice sector was technologically backward and plagued by onerous credit and marketing relations. It is very likely that a considerable proportion of these large-farm operators were market intermediaries as well (D. Wong 1983). Ten of my final sample of thirty-three large-farm operators in 1985 were rice buyers while another was a shopkeeper.

The favorable change of policy towards the rice sector (see Lim and Muhammad Ikmal Said, Chapter 9) has naturally strengthened wealthy farmers. Vast improvements in drainage and irrigation and institutional infrastructure have facilitated the narrowing of the "productivity gap" which once separated them from the poor peasantry (De Koninck 1983:38; Lim 1985:32). Large-farm operators are also the primary beneficiaries of state fertilizer and price subsidies.

Under the fertilizer subsidy scheme, farmers are entitled to two bags of urea, four bags of mixed fertilizer, and one bag of ammorphos for every 0.4 hectare they cultivate. A farmer can claim up to a maximum of 2.4 hectares only. However, all operators of large farms have no problems claiming subsidized fertilizer up to their actual farm size. As for the price subsidy, farmers are reimbursed M$16.54 for every 100 kilo of unhusked rice sold at any approved rice mill. Large farms in my 1985 study often produce between 4.7 to 4.9 tonnes per hectare. Thus, a 10-hectare farm operator receives a subsidy of between M$7773.8 and M$8190.6 per season. The largest six farms in the 1985 study operate more than 28.77 hectares. This means that they receive more than M$22,300 per season or M$44,600 a year from the rice price subsidy alone.

Double-cropping has necessitated mechanization of the plowing and harvesting processes. Wealthy farmers have gained considerably by investing and contracting out their tractors and combine harvesters (Muhammad Ikmal Said 1985a; Mustafa 1985). Mechanization has also partially removed labor constraints (in terms of supply and high wage rates) that once hindered farm expansion. This seems to explain why a considerable proportion of large farms increased their size during those periods when mechanization was introduced (Muhammad Ikmal Said 1985a:58–59).

Mechanisms of Access to Land

Access to land is secured predominantly by inheritance, purchase, and renting land owned by others. The rapid development of productive forces, particularly the successful implementation of the Green Revolution in the 1970's and 1980's, has resulted in increased competition for land, driving up land prices and rents. The price of land increased fourfold between 1970 and 1980 (Muhammad Ikmal Said 1985a:93–95). No small operator can afford to purchase land that presently costs about M$35,000 per hectare.

Land rents have also increased, but at a rate well below increases in the price of land. As a result, absentee landlords are increasingly faced with a situation where returns to land are on the decline (Muhammad Ikmal Said 1985a:70). This explains the apparent tendency for absentee landlords to sell off their holdings. I encountered one such case in my 1981 study and another in 1985. The 1981 case involved the sale of an unusually large lot of 69.9 hectares by a moneylending family concern to a group of eleven wealthy Chinese businessmen-farmers. This land was sold at M$22,935 per hectare. In 1985, a 43-hectare lot was said to have been sold to a private company for M$1.6 million (or M$37,066 per hectare). To the best of my knowledge, this is the first time in the Muda area that a private nonfamily corporation has invested heavily in land with the intention of organizing rice production directly. Thus, although wealthy capitalist-farm entrepreneurs are also affected adversely by increases in the price of land, they have been able to mobilize large sums of capital to purchase land. In the process, tenants have been marginalized; seventeen tenants in the first case and twelve in the second were paid off to make way for these new "masters of the countryside."

The inability of poor farmers to secure sufficient land is also related to the competition for land offered for lease. It is obvious that rich capitalist farmers hold a domineering advantage in this competition. There are at least two ways in which large capitalist farms outbid their poorer counterparts. First, they are able to offer relatively higher rents for a given unit area. A recent comparative study of large and small farms by Lim (1985:29) shows that large farms pay rents at levels that are almost twice those paid by small farms.[1] Second, rich farmers can also outbid others by offering landowners longer contracts.

In addition, straitened landowners (particularly small ones) sometimes borrow money from rich farmers. In exchange, these landowners transfer the usufructuary right over their lands to the creditors for a certain period. The period of these "usufructuary loans" (Wilson 1958:16) depends on the amount of loan taken as well as the negotiated equivalent rent their lands fetch. Usufructuary loans are an analytically different form of lease because land surrendered to the creditors is not offered in the land-lease market. Thus, usufructuary loans reflect two converging class trends: on the one hand, pressing economic circumstances impel straitened owners to exchange their land for credit; on the other, "agriculturalist moneylenders" (Ghosh and Dutt 1977:174) often lend money in order to secure more land for cultivation. The extent of usufructuary loans is not easily detectable because usufructuary loans may pass easily as different forms of cash advances for leasehold agreements.[2] At least six of the thirty-four rich farmers I studied in 1985 were known to lend money to their clientele. One of them claimed that his clients borrow (for both consumption and production purposes) up to about M$70,000 per season. This does not suggest that poor peasants usually surrender their land for credit, but some large farmers can exert influence over their clientele to provide access to that facility.

The different strategies of securing land are important because rented-in land forms a high proportion of large farms. In 1981, rented land made up about 70 percent of the area of large farms (Muhammad Ikmal Said 1985a:69). This has been corroborated further by my recent study; 66 percent of the total farm area of the thirty-three farms studied in 1985 was leased. Leasing for more than two-season periods is the main mode through which large Chinese farms secure land. Long leases are less common among Malay farmers. Even so, the predominant mode of farm expansion among Malay farmers was through renting land (predominantly through seasonal rents) owned by others (Muhammad Ikmal Said 1985a:72–73). The availability of land for rent thus facilitates the expansion of capital in rice cultivation and enables the owners of capital "to circumvent one of the fundamental contradictions of capitalist agriculture, that of the burden of land ownership" (De Koninck 1981:31).

A few of the Chinese farmers showed keen interest in rehabilitating hundreds of hectares of abandoned riceland, at low rents in the neighboring state of Penang. These wealthy farmers are thus considering cultivating rice on a much larger scale. They are aware that the state encourages the rehabilitation of these lands. However, a large part of these lands are owned by Malays, a backward minority in the state of Penang. The current ethnic situation may, therefore, inhibit their capacity to expand.

Inheritance is an important mechanism through which poor peasants get access to land (Fatimah 1980:406, 416; D. Wong 1983:128). Although most landless peasants are claimants of inheritable property, the amount of inheritable land is often very small (Mohd. Shadli 1978:76). Members of large-farm households often inherit not only larger areas of land but also capital (Muhammad Ikmal Said 1985a:28–40; D. Wong 1983). However, the manner in which farm and nonfarm capital is reproduced intergenerationally differs between Chinese and Malay operators.

Inheritance and the Internal Organization of the Household Economy

Because of different inheritance systems, the Chinese family-farm enterprise has remained intact over a longer period than its Malay counterpart. Among the Chinese, the division of property is usually administered before the death of the father-owner and male children receive approximately equal shares. Married male siblings often reside under the same roof and cultivate the family farm jointly under the direction and supervision of the family patriarch. This relatively early division of the family patrimony ensures a definite, if not an equal, return to the labor contribution of each son under the father's or eldest brother's management of the household farm, and it provides the infrastructural basis for Chinese households to operate a single farm jointly over a fairly long period. The long duration of joint operation facilitates accumulation. When the property of such enterprises is eventually divided, the splintered farms are relatively large and their owners well placed to rent additional land.

In contrast, Malay family farms disintegrate rapidly with marriage of children, who form separate and relatively independent households. Formation of these new domestic units is often aided by land and other resources from the parents' enterprise (Mohd. Shadli 1978:57–60). Land provided by parents to splinter units is not given out permanently, as it

TABLE 11.1. *Inheritance and Farm Size over the Developmental Cycle*

	Marriage	Point of entry[a]	1975	1985
Chinese (*n* = 12)[b]				
Age	23	34	46	56
Land inherited (ha)[c]	0.36	4.53	5.27	6.33
Farm size (ha)[d]	12.06	11.27	18.21	24.99
Malays (*n* = 7)[b]				
Age	24	21	39	49
Land inherited (ha)	0.12	0.66	2.06	2.47
Farm size (ha)[d]	5.02	2.84	10.30	12.04

[a] "Point of entry" is defined as the point at which a farmer first assumes control and management of a farm. This occurs either when a person breaks away from the family unit to form a new farm or when a head of household retires or dies.
[b] Includes only those farmers whose parents' estates have been subdivided.
[c] Includes 15.83 ha of rubber land inherited by two individuals.
[d] Rice farm size only.

is often passed on to other siblings upon their marriage or reclaimed by parents (Mohd. Shadli 1978:60). Therefore, although inheritance in Malay family farms is usually executed after the death of parents (Kuchiba et al. 1979:52), the process of fission occurs much earlier, quite soon after marriage of the first child. As a result, farm expansion is checked.

This difference between Chinese and Malay farms partly explains why Malay farmers begin to farm independently at an earlier age, receive smaller amounts of inheritance, and cultivate comparably smaller farms with smaller households that break up gradually with marriage of their children (Muhammad Ikmal Said 1985a:42–47). The twelve Chinese households in my sample in which land transmission had taken place established their farms when they were about thirty-four years old; in contrast, the seven Malay farmers formed their own farms at an average age of twenty-one, three years before they got married. This group of Chinese farmers generally owned more land, operated much larger farms, and inherited larger amounts of land (Table 11.1). Their parents also owned larger holdings. Parents of the Chinese farmers owned an average of 16.2 hectares, compared with 9.5 hectares for the parents of Malay farmers.

The ability of Chinese farms to sustain their scale over time may also be attributed to the fact that their farm enterprises are more extensively backed by other forms of business. My evidence suggests that Chinese farmers have trodden two paths of diversification. The first group diversified into rice production after having established themselves in (rice-related) trade, while the second diversified into other (but mainly rice-related) sectors of the economy from paddy farming. In general, the first group has been more successful, partly because their trading-cum-moneylending activities in the past placed them in a very strategic position.[3] Today they have investments that integrate both the downstream and upstream activities, such as retailing rice and farm inputs, machine (tractor and combine harvester) contracting, and rice milling. A few have moved some of their capital out of the rice sector altogether, with investments in real estate and other businesses. However, regardless of their family origins (i.e., whether they were businessmen, farmers, or both), rich Chinese farmers have enhanced their present status in farming by their participation in other business undertakings, and the organizational structure of the family provides it with the capability in terms of labor and capital to venture into business over a fairly long period.

The parents of the wealthy Malay farmers were all farmers. Although seven of the nine Malay farmers have diversified into other areas, hardly any have done so in a persistent manner and on a scale that is comparable with the Chinese farmers. Apart from the generally smaller quantum of wealth their parents owned and the manner such wealth was divided, the small-scale and temporary nature of their businesses may also be related to the manner in which their household economy is organized. Thus, Malay families do not have a structural mechanism which encourages their children to stay with their parents and operate businesses jointly. Early and gradual fission of the family, as well as the practice of withdrawing some of the parents' own resources for newly married children, makes it difficult for parents of Malay farming families to build and sustain a business venture over a relatively long period. Nonfarm operations, particularly, are not physically divisible; it is therefore difficult to sustain an investment that keeps declining (due to shortage of family labor and funds) with the development of the family. As discussed below, some Malay farmers in my sample are involved in other businesses. However, household organization and inheritance systems undermine their capacity to build and sustain a business enterprise.

Political Linkages and Access to Resources

Studies of village communities in Malaysia (Kessler 1978; Scott 1985; Shamsul 1986; Syed Husin 1975) consistently point out that ownership of means of production and political influence are mutually reinforcing. Mansor's study (1982) on large landowners in Kedah reveals similarly that they wield considerable political power both at village and extra-village levels. This mutual reinforcement between economic and political power in the Malaysian countryside is particularly evident in two recent studies by Scott (1985) and Shamsul (1986). Shamsul shows clearly how in Malawati, a rural district in the state of Selangor, politicians collaborate closely with influential village leaders in the distribution of development expenditure at the district level. Similarly in Sedaka, a village in the irrigated Muda area, UMNO leaders at the village level are comprised of wealthy peasants, who, by virtue of the mutually reinforcing influence of their economic and political power, control the Farmers' Association (FA) and other state institutions, such as the Village Development Committee (Scott 1985:125–137). As a result, the distribution of credit (the FA) and other handouts (such as the Village Improvement Scheme program) are often restricted to them and their supporters.

There are few Chinese villages in the Muda area, and large-farm Chinese operators generally reside along the main roads that pass through their districts or *mukims* (smaller administrative units of a district). Although wealthy Chinese farmers do not enjoy the forms of benefits their Malay counterparts get through control of the village development machinery, these Chinese farmers and party leaders are influential within the Chinese and Malay communities. This influence is due not only to their considerable wealth, but also because the Chinese community forms a politically decisive voting bloc in some state and parliamentary constituencies in Kedah where Malay voters are divided quite evenly between PAS and UMNO. My informal conversation with the Chinese large-farm operators revealed that at least four of the twenty-five are office-bearers of Chinese-based component parties of the National Front (or Barisan Nasional, the ruling alliance); of these four, three received commendatory awards from the Sultan of Kedah and even the Agung, Malaysia's head of state.

Most Chinese farmers are not members of the FA, although no formal criteria for membership discriminate against them. Many complained that their recent application for membership was turned down on grounds that the FA has interests in Malay Land Reservations (and perhaps other *bumiputra*-specific, i.e., Malay-specific investments). Unconvinced, they argue that *bumiputra* interests of the FA could be separated from its other functions. A check with an official of the Muda Agricultural Development Authority revealed, as the large-farm Chinese entrepreneurs also realize, that there are Chinese farmers who are members of the association. It was also related by the official that it is only recently that rich Chinese farmers are showing interest in the FA. Their lack of interest earlier was undoubtedly related to the uncertainty of the FA to get its feet off the ground. In 1974, total membership in the FA was 11,693 while its share capital amounted to M$241,245. By 1984, the FA had 33,613 members and its total share capital increased to M$4.2 million (Abu Hassan 1985). Clearly, rich Chinese farmers in the Muda area are trying to gain access to the FA's increasing capital and investments—particularly, by their own admission, to the cheap credit and other facilities (for example, cheaper transport costs for delivering subsidized fertilizer) made available through the FA. Another complaint related to their unsuccessful application for membership into the FA is that they do not have any say in decisions that affect their operations. It is perhaps significant that they are beginning to show keen interest in the FA after it has grown considerably while simultaneously fully conscious of their political position in the PAS–UMNO rivalry. Very little is known of the "investment" in politics by the very wealthy but politically minor Chinese farmers. However, the political links they might have forged with the rural Malay elite must necessarily form part of their process of capital accumulation, which is increasingly spread over various forms of investment in the countryside.

Diversification and Relations of Indebtedness

Of the thirty farmers studied in 1981, eleven were involved solely in rice cultivation. Of the remaining farmers, ten were involved in both rice cultivation and agriculture-related activities, six in rice cultivation and non-agriculture-related activities, while three others were found to be involved in all the three types of activities mentioned above (Lim 1985:11).[4] The diverse economic interests of large farms were confirmed further in my 1985 study. Nine of the thirty-three farmers (seven Chinese

TABLE 11.2. *Diversification by Large Farmers*

Type of investment	No. of Chinese	No. of Malays	Total
A. Rice production only	7	2	9
B. Rice production and other agriculture-related activities: Rice trading (5) Rice milling (3) Other trade (4) Landownership (3) Rubber production (5) Combine harvester contracting (6)	11	6	17
C. Rice production and nonagriculture	1	0	1
B and C both	5	1	6
Total	24	9	33

and two Malay) had interests limited to rice production. Another seventeen had investments in rice production and other agriculture-related activities (including rice trading, combine harvester contracting, rice milling, rubber production, and landownership—i.e., mixed farming, riceland). Seven farmers had interests in rice production and non-agriculture-related activities (such as petty trading, housing development, transport), and six of these were also involved in agriculture-related activities. This information is summarized in Table 11.2.

These merchant-farmers actively try to establish links between their marketing (and machine-contracting) functions and moneylending. There is, therefore, a historical continuity of the close relationship between marketing and moneylending. The need for credit compels poor farmers to become clients of these merchant-farmers. From the merchant-farmers' point of view, the provision of credit ensures a clientele for their services; farmers are obliged to sell their produce in order to settle their debts with merchant-farmers who provide them credit. This obligation is necessitated by the expected need for further credit in the next round of production. As I mentioned earlier, there were at least six rich farmers who provide credit for their clientele.[5] By doing so, they form an almost captive market for the goods and services provided by the wealthy farmers. Therefore, the cycle of peasant reproduction forms an important link in the reproduction of capital.

Indications of the importance of moneylending to machine contractor-farmers emerged in the course of field research. One farmer who contracted his tractor to other farmers related that it is a profitable undertaking. However, payment for the tractor hire was often very slow because farmers are generally poor. In other words, peasants who hired this farmer's tractor services often failed to pay and were, therefore, indebted. As security of payment, the indebted peasants were required to sell their produce to this tractor contractor. This particular farmer has sold off his tractor to a friend, and it is this friend who now buys their produce. This description emphasizes the relationship between delayed payments of tractor services by poor peasants and the sale of their produce through merchant-farmers. It is plausible that delayed payment is an undisclosed marketing ploy; more peasants would hire their services when terms of payment are easier.

The search for clients is even more pressing when the amount of investment involved is large. Thus, investment in combine harvesters (which cost about M$190,000 in 1980, M$240,000 in 1985) means that its owners have to "spread their wings" beyond their traditional geographical areas. The appointment and emergence of brokers was a response to this development. Brokers are influential (sometimes "feared") local residents who secure clients for owners of combine harvesters. The appointment of such agents is also necessitated by the need for consent from farmers for free passage through their fields. Under these circumstances, it can be expected that owners of combine harvesters strengthen their preexisting relations (services) or create new ones as individual production is socialized under their direction. Nine of the thirty-three large farm units studied in 1985 owned twenty-five combine harvesters, six of which were wholly owned.

Mechanization, Accumulation, and Changing Employment Patterns

Investment in farm machinery hardly existed in the sixties. In 1961 only about 5597 hectares (5.8 percent) of the Muda Plain was plowed mechanically (Jegatheesan 1980:5). By 1974, some 92 percent of land preparation was mechanized in the off season and 76 percent in the main season (Yamashita et al. 1981:29). The rapid mechanization of the plowing process is undoubtedly related to the change in state policy towards the rice sector and the concomitant development of the productive forces within it.

The increase in yields amidst the wide range of state protection and support (Lim and Muhammad Ikmal Said, Chapter 9) resulted in a general increase in farm incomes which, in turn, led to the general proliferation of market outlets. These developments relieved the peasantry of the harsh burdens of usury and relatively high rents (in terms of its proportion to the total cost of production) that characterized the paddy sector before the sixties (Muhammad Ikmal Said 1982). Consequently, less capital was tied up in usury and owners of capital (including farmers who also had interests in moneylending and rice trading) and the peasantry were increasingly able to devote more capital to production. As the number of tractors increased, the cost of hiring them declined. This catalyzed further the process of mechanization in the rice sector.

Investment in farm machinery was initially undertaken primarily by businessmen-contractors (Jegatheesan 1980; Malaya 1956:125). It is only recently that operators of large farms have begun to invest in farm machinery. Only three of twenty farms (from a total of twenty-eight large farms I studied in 1981) that had come into existence in 1965 owned any tractor, and these were confined to Malay farmers. Jegatheesan (1981:3) suggests that investment in farm machinery by operators of large farms became apparent only in the early seventies. Investment in tractors by large farms in my 1981 study confirms this; by 1975, fifteen out of twenty-five farms that had come into existence owned twenty-one units. Eleven of these units were two-wheel tractors while ten were four-wheel tractors.

Operators of large farms have since invested in combine harvesters. Nine of the thirty-three wealthy farmers (all Chinese) studied in 1985 owned twenty combine harvesters, six of which were wholly owned. Six of the nine combine harvester owners (with partial or full control over seventeen units) are found among "Group B" farmers (see Table 11.2). Three others from this group had just sold their combine harvesters after operating them for some years.

Investment in farm machinery augments the position of large farms. Owners of combine harvesters achieved a dramatic increase of about 76 percent of their 1981 farm area between 1975 and 1981. As I have argued more fully elsewhere (Muhammad Ikmal Said 1985a), investment in combine harvesters has facilitated the expansion of large farms in the Muda area. Profits from combine-harvester contracting were especially large when combine harvesters were few in number. In 1977, the average net profit from each combine harvester was M$41,600 per season (or M$83,200 per year). However, by the second season of 1980, the average net profit declined to about M$29,500 and decreased further to about M$21,000 in the first season of 1981 (Mustafa 1985:32). The average rate of profit for 1981 was estimated to be lower than the prevailing lending rate. As a result, several owners have forfeited or sold off their machines.

Operators of large farms who do not own combine harvesters have also benefited greatly from the mechanization of rice harvesting because of the significant reduction in the cost of the most labor-intensive part of the production process. Manual reaping presently costs about M$156 per hectare while manual threshing costs about M$186 (assuming a yield of about sixty-two sacks per hectare at M$3 per sack). The combined cost of manual harvesting is, therefore, M$342 per hectare. On the other hand, combine harvesting costs only M$208.50 per hectare. Thus the twin problem of increased labor costs and labor shortage has been resolved in a way that strengthens further the position of large farms.

The immediate effect of the introduction of combine harvesting is the displacement of labor. A study of the ACRBD 4 (irrigation block area) in Locality D2 of the Muda area in 1979 reported a decrease of 65 percent of the total labor input between the off-season of 1978 and the main season of 1979 (Yamashita et al. 1981:38). An estimated total of 10,800 man-days would have been lost should the whole of the ACRBD area be combine-harvested. This would mean that a total of 514 persons would be displaced if harvesting time were limited to three weeks (Yamashita et al. 1981:38). It is estimated that about 68 percent of poor households that were once involved in harvesting work have been displaced (Mustafa 1985:40).

Whether the displacement of labor in the harvesting process has resulted in declining incomes is a matter of dispute. While there is no doubt that the drastic decline of employment opportunities connected with reaping the crop manually has meant the loss of income (Barnard 1981:221-224; Scott 1983:73; Yamashita et al. 1981:138), Mustafa (1985) maintains that these short-falls have been more than compensated by increased employment opportunities (such as bagging unhusked rice, transporting the harvest from the field to the bunds, as well as other farmwork such as fertilizing the fields) and increased wage rates (due to labor scarcity, a phenomenenon yet to be verified and explained scientifically). Mustafa also reports that although 68 percent of small-farm households were displaced from harvesting work between 1972 and 1982, 64 percent of them experienced an increase in real income from working on other people's farms. More research is needed to verify this surprisingly high percentage of people who are claimed to secure higher incomes in spite of their displacement from the harvesting process.[6]

Unlike harvesting, transplanting of young shoots from nurseries for cultivation in the field is still, to a considerable extent, dependent on manual labor. The cost of transplanting labor, undertaken almost exclusively by women, has, like other forms of labor, increased tremendously. It has risen from about M$87 in 1972 to M$226 per hectare in 1985, although there is hardly any increase in real terms.[7] Farmers seeking to avoid the cost increase have resorted to seed-broadcasting, which requires about one-third of the labor needed for transplanting. About 12,000 hectares (or 12.6 percent) of the Muda rice-growing area was seed-broadcasted in 1982. However, operators of large farms do not generally resort to seed-broadcasting because the risks are higher (inundation due to heavy rain and heavy weed growth greatly reduce the yield). Six of the thirty-three operators of large farms studied in 1985 reported using labor from southern Thailand. Transplanters from southern Thailand are cheaper as they charge about M$156 per hectare. They probably cost about M$208 per hectare when food and transport costs are included. Apart from the fact that they are cheaper, laborers imported from southern Thailand also form a dependable source since they are available during the period they are needed. Respondents claimed that local village laborers are less predictable (because they have to attend to other farmers' plots) even when they have received advanced payment for transplanting.

It appears, therefore, that the development of capitalism in the Muda area has generally increased the purchasing power of its households precisely because large farms (not necessarily confined to those beyond 10 hectares) have strengthened their position. My data suggest that large farms are operated by a few individuals who assume merely managerial/supervisory roles with increasing interests in economic activities not directly related to rice cultivation (such as combine-harvester contracting, trading, etc.). This possible trend of decreasing participation in farm work by large-farm households (while farm size is increasing) perhaps explains the increase in farmwork for households of smaller farms. Their increasing investment in activities not directly related to rice production also generates alternative employment (such as the crew for combine harvesters, labor in rice mills, etc.), supporting to some extent the development of the "home market."

Conclusions

The discussion above has highlighted large-farm strategies of capital accumulation and the possible implications these have on the peasantry, particularly among the poorer classes. Mechanisms of accumulation include the system of inheritance and the attendant organization of family resources and political alliances with ruling elites, which in turn shape diversification of investment in agricultural (particularly rice-related) and nonagricultural business undertakings.

Although operators of large farms constitute a very small minority, the evidence suggests that their influence (both as farmers and as businessmen-farmers) extends well beyond their number. As farmers, they are the largest employers of wage labor and the largest tenants of land. Their influence in the rice economy is even more extensive when they also have investments in rice-related business ventures such as rice trading, rice milling, farm inputs retail trading, sundry-goods retailing, moneylending, and machine contracting. This is because such investments bring them into contact with a large number of farmers. Of greater importance is the fact that the services they cater for the farming community are linked integrally to the production-reproduction cycle of the peasantry. As a result, the influence of large-farm operators (a little more than two-thirds of whom are involved in other forms of investment) is both extensive and critical to our understanding of "peasant society." For example, the provision of credit enables reproduction of the poor peasantry who in turn provide the market for goods and services provided by operators of large farms.

The theoretical importance of this point is that attempts to understand the dynamics of "peasant society" must take into consideration the existence of the small minority of wealthy farmers, even when the proportion of small farms appears to be on the increase (as in the case of the Muda area). A great deal of insight is lost by focusing narrowly on the internal characteristics of small farms, because small farms are considerably dependent on large farms for their reproduction.

Notes

1. Evidence on the relationship between farm size and rent levels is conflicting. Wong Hin Soon's study (1983:113, 115) shows that this is true only for owner-tenants. This uncertainty may be due to the fact that aggregated rent level figures do not distinguish between different types of rent agreements. This difference is important because poor farmers usually do not prefer leasehold agreements as they involve large sums of money. Second, the level of rents under leasehold agreements (tenure agreements that span over two or more seasons) are generally higher than rents that are paid seasonally (Gibbons et al. 1981:80). However, it needs to be stressed that rent levels under leasehold agreements are higher than seasonal rents only for the smaller land parcels. Leasehold rents for large land parcels are well below rents paid seasonally for land parcels of similar size.

2. Mohd. Shadli (1978:95–96), for instance, acknowledges that one of the major reasons for renting out land is the need for cash. However, this coterminous credit-tenurial arrangement is seen as just another form of lease (Mohd. Shadli 1978:107–108). Others (Mokhzani 1973:289; Wilson 1958:16), however, treat this arrangement as a form of credit relationship that is mediated through usufructuary rights over land.

3. Before the 1960s, moneylending was highly profitable as the interest rates charged then were very high (Mokhzani 1973; Thompson 1954). By borrowing money, straitened peasants were often caught in the vicious cycle of indebtedness, sometimes impelling the sale of portions of their land to the moneylender-trader. In addition, indebted peasants are also often obliged to sell their paddy to them at low predetermined prices. In other instances, loans were given out in return for usufructuary rights to the debtors' lands (Wilson 1958). These tied-in conditions of exchange provided these farmer-traders direct access to (either to buy or rent) their clients' lands. For example, the operator of the largest known farm in Muda has in his possession the ownership grant of a 2.88-hectare lot which he has "rented" (since 1966) under some form of credit arrangement.

4. These two studies on large farms shared the same opening inventory data. With the exception of a few individuals, respondents in the two studies were the same.

5. I am inclined to think that a large majority of the wealthy farmers who perform at least two roles provide credit to their poorer counterparts. One rich Malay farmer estimated that about 75 percent of farmers who farm along the same canal bund are indebted. Many of these loans were for consumption purposes. I noticed on a number of occasions during my interviews that the money borrowed was of small amounts—amounts too small to constitute payment for a factor of production. One creditor-farmer remarked that some even borrow to buy milk for their children.

6. The impact of mechanization on income would be gauged more accurately if the base year corresponded to the period just before combine harvesting spread throughout the region in 1978–1979. Syed Ahmad (1984, 1986) reports that average income levels declined and the incidence of poverty increased after 1975. This corroborates Scott's findings (1985:113–120) in which 1977 was used as the base year.

7. The cost of transplanting was about M$121.60 in 1978–1979, or almost exactly half of the present cost (Barnard 1981:220; Jegatheesan 1981:5).

References

Abu Hassan Md. Isa (1985). "Penilaian Prestasi Pergerakan Pertubuhan-Pertubuhan Peladang MADA: Aspek Ekonomi." In *Seminar Peladang MADA, 1985: Kertaskerja dan Temuan*. Alor Setar: MADA.

Afifuddin Hj. Omar (1978). *Peasants, Institutions and Development in Malaysia: The Political Economy of Development in the Muda Region*. MADA Monograph no. 36. Alor Setar: Muda Agricultural Development Authority.

Barnard, R. (1981). "Recent Developments in Agricultural Employment in a Kedah Rice-growing Village." *Developing Economies* 19:207–228.

Berwick, E.J.H. (1956). *Census of Padi Planters in Kedah, 1955*. Alor Setar: Department of Agriculture.

Bray, F. (1983). "Patterns of Evolution of Rice-growing Societies." *Journal of Peasant Studies* 11:3–33.

De Koninck, R. (1979). "The Integration of the Peasantry: Examples from Malaysia and Indonesia." *Pacific Affairs* 52: 265–293.

——— (1981). "Of Rice, Men, Women and Machines in Malaysia." *Journal Ekonomi Malaysia* 3/4:20–37.

——— (1983). "Getting Them to Work Profitably: How the Small Peasants Help the Large Ones, the State and Capital." *Bulletin of Concerned Asian Scholars* 15:32–41.

——— (1985). "On the Usefulness of the Income-sharing Thesis." *Kajian Malaysia* 2:36–54.

Fatimah Halim (1980). "Differentiation of the Peasantry: A Study of the Rural Communities in West Malaysia." *Journal of Contemporary Asia* 10:400–422.

Ghosh, A., and K. Dutt (1977). *Development of Capitalist Relations in Agriculture*. New Delhi: People's Publishing House.

Gibbons, D. S. (1984). *Paddy Poverty and Public Policy*. CPR Monograph Series no. 7. Penang: Universiti Sains Malaysia.

Gibbons, D. S., et al. (1981). *Land Tenure in the Muda Irrigation Area: Final Report, Part 2: Findings*. CPR Monograph Series no. 5. Penang: Universiti Sains Malaysia.

Jegatheesan, S. (1980). "Progress and Problems of Rice Mechanisation in Peninsular Malaysia." Paper presented at the Presidangan Padi Kebangsaan, Malaysia, Kuala Lumpur, 26–28 February.

——— (1981). "Progress and Present Status of Mechanisation in Muda." Unpublished mimeo. Alor Setar: MADA.

Kahn, J. (1980). *Minangkabau Social Formations: Indonesian Peasants and the World-economy.* Cambridge, Cambridge University Press.

——— (1982). "From Peasants to Petty Commodity Production in Southeast Asia." *Bulletin of Concerned Asian Scholars*, vol. 14, no. 1, 3–15.

Kessler, C. S. (1978). *Islam and Politics in a Malay State: Kelantan 1838–1969.* Ithaca: Cornell University Press.

Kuchiba, M., et al. (1979). *Three Malay Villages: A Sociology of Padi Growers in Malaysia.* Honolulu: University Press of Hawaii.

Lim Teck Ghee (1985). *Small and Large Farms in Muda: A Comparison of Organization of Production, Yields and Profitability.* CPR Monograph Series no. 9. Penang: Universiti Sains Malaysia.

Maeda, K. (1967). *Alor Janggus: A Chinese Community in Malaya.* Kyoto: Center of Southeast Asian Studies, Kyoto University.

Malaya (1956). *Final Report of Rice Committee, 1955.* Kuala Lumpur: Government Printers.

——— (1961). *Census of Agriculture, Federation of Malaya: Preliminary Report No. 3 ("Type, Tenure and Fragmentation of Farms").* Kuala Lumpur: Ministry of Agriculture.

Mansor Hj. Othman (1982). "Political Power and Wealth in the Rural Areas: A Case Study of the Padi Landed Elite in Kedah." *Akademika* 20/21: 374–394.

Mohd. Shadli Abdullah (1978). "The Relationship of the Kinship System to Land Tenure: A Case Study of Kampung Gelung Rambai." Unpublished M.Soc.Sc. thesis, Penang, Universiti Sains Malaysia.

Mokhzani, B.A.R. (1973). "Credit in a Malay Peasant Economy." Unpublished Ph.D. thesis, University of London.

Muhammad Ikmal Said (1982). "Commoditisation and Forms of Capital Accumulation in the Padi Sector, West Malaysia." *Akademika* 20/21:327–355.

——— (1985a). *The Evolution of Large Paddy Farms in the Muda Area.* CPR Monograph Series no. 8. Penang: Universiti Sains Malaysia.

——— (1985b). "Household Organization of Capitalist Farms and Capitalist Development in Agriculture." *Kajian Malaysia* 3:101–128.

Mustafa Mohd. Najimuddin (1985). *The Political Economy of Mechanization: The Case of Combine Harvesters in the Muda area, Kedah.* CPR Monograph Series no. 10. Penang: Universiti Sains Malaysia.

Scott, J. (1983). "Api Kecil dalam Pertentagan Kelas." *Kajian Malaysia* 1:62–94.

——— (1985). *Weapons of the Weak: Everyday Forms of Peasant Resistance.* New Haven: Yale University Press.

Shamsul, A. B. (1986). *From British to Bumiputra Rule: Local Politics and Rural Development in Peninsular Malaysia.* Singapore: Institute of Southeast Asian Studies.

Syed Ahmad Almahdali (1986). "Status and Problems of Padi Production in the Muda Irrigation Project," Paper presented at the Persidangan Padi Kebangsaan, 1986, Kuala Lumpur, 20–22 July.

Syed Ahmad Almahdali and Baharuddin Mahmood (1984). "Penglibatan Lembaga Kemajuan Pertanian Muda Di Peringkat Sebelum Tuai Dalam Industri Padi Kedah/Perlis, Kawasan Serta Perbelanjaan Terlibat Bagi Memajukan Industri Ini." Paper presented at the *Kongress Penyertaan Orang Melayu Dalam Industri Padi Dan Beras*, Sungai Petani, 12–14 April.

Syed Husin Ali (1975). *Malay Peasant Society and Leadership.* Kuala Lumpur: Oxford University Press.

Thompson, A. P. (1954). *Report on the Marketing of Rice in the Federation of Malaya.* Report no. 278. Rome: FAO.

Wilson, T. B. (1958). *The Economics of Padi Production in North Malaya.* Bulletin no. 103. Kuala Lumpur: Ministry of Agriculture.

Wong, D. (1983). "The Social Organization of Peasant Reproduction: A Village in Kedah." Unpublished Ph.D. thesis, University of Bielefeld.

Wong, Hin Soon (1983). *Muda II Evaluation Survey—An Impact Evaluation Study of the Muda II Irrigation Project.* Alor Setar: Muda Agricultural Development Authority.

Yamashita, M., et al., (1981). *Farm Management Studies. MADA–TARC Cooperative Study, Pilot Project ACBRD 4, Muda Irrigation Scheme.* Technical Bulletin no. 14. Ibaraki, Japan: Tropical Agricultural Research Center, Ministry of Agriculture Forestry and Fisheries.

Part Five

Indonesia

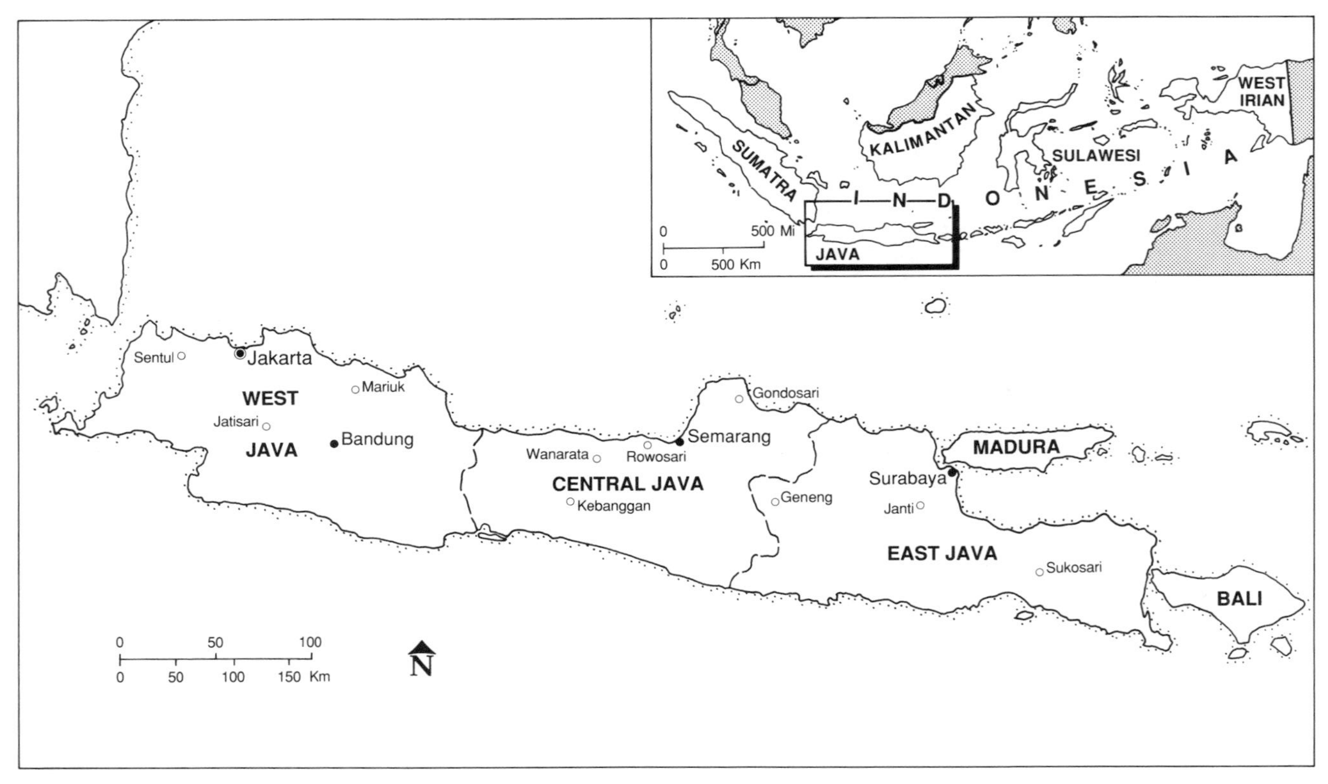

Map 4. Indonesia and Java

Chapter Twelve

Java: Social Differentiation, Food Production, and Agrarian Control

Frans Hüsken and Benjamin White

The fourteenth-century Javanese Prince of Wengker was well aware of the critical importance of the peasantry and peasant production for the security and cosmic balance of his realm. Addressing his subjects, he formulated quite clearly what he wanted them to be: orderly, submissive, and industrious cultivators:

> O you, do not be untrue, be most faithful, submissive, loving, Protector-obeying [loyal] toward our lord. Be orderly, you in the third estate [*waishya*]. Anything that might cause welfare of the rural districts [*pradeshas*], that should be had at heart. Dams, dike-roads, *wandiras* [Ficus Rumphii], *gerhas* [buildings] etcetera, all kinds of those good *kirtis* [foundations] should be kept in good state. The principal of it: *gagas* [unterraced fields], *sawahs* [terraced fields], anything that is planted: that it may thrive, it must be guarded and treated with care. . . . The becoming greater of the *pradeshas* there, that is what should be striven after.
>
> (*Nagara-kertagama* Canto 88, Stanzas 2 and 3; Pigeaud 1960:III, 103–104)

Six centuries later, a similar view of the peasantry still imbues Indonesian government agricultural and rural development policy although rural conditions have drastically changed.

Once a sparsely populated region, Indonesia is now the most populous of the ASEAN nations, with two-thirds of its people crowded on only about 7% of its land area in the island of Java, whose population densities are higher than any other agrarian region of comparable size in the world.

Our main purpose in this paper is to examine the forces which have influenced patterns of social-economic differentiation under conditions of acute population pressure in rural Java both before and since Indonesian independence and particularly during the years of the 'New Order' regime (1966–present), which more or less coincide with the Green Revolution in rice cultivation.

Many early studies of the consequences of the Green Revolution in Java have pointed to the unequal distribution of the benefits of the new technologies and have identified various mechanisms by which this has occurred. The larger farmers—however small their holdings in absolute or comparative terms, as we will see below—more quickly adopted the new technologies, having greater access to subsidized credit and other inputs (Bryant 1973:292–313; Franke 1972; Soewardi 1976). They also reduced labor costs (the largest single item in production costs) through a variety of 'rationalizations' of the labor process, some of which will be described below and in Chapters 13 and 14. Summarizing many of these reports in the mid-1970s, Palmer concluded that commercial and technological innovations in Javanese rice production had 'provoked new forms of social differentiation resulting in schisms in relations between (a) farmers, on the one hand, and landless laborers and very small owner-cultivators on the other, and (b) between those landless laborers who are offered the limited work at any time and those who are unable to share in job opportunities' (Palmer 1976:24).

Many researchers consider these developments as proof that the conventional view of Java as a relatively egalitarian and homogeneous society is no longer valid. The transformation of the rural structure and particularly the emergence of a commercialized peasant agriculture are seen as marking the beginnings of social differentiation, the end of 'involution,' or even as the inception of rural capitalism in Java (Collier 1981a; Gordon 1978). Such interpretations involve a too-easy acceptance of prevailing views of 'traditional' Javanese society and in turn a misleading interpretation of the changes occurring during the past two decades of political-economic reorganization and production intensification, as if social differentation and commercialization in the peasant sector were very recent developments. Against that view, a countercurrent has (in our view, more correctly) seen Javanese society as historically divided into agrarian classes based primarily on differential access to land and characterized by a long but uneven history of commercialization, so that in turn the changes associated with the recent Green Revolution are better seen as the continuation, crystallization, or reemergence of trends set in motion long ago.

After discussing these problems in the characterization of Javanese rural society in the colonial period, we will consider the various political imperatives confronting the 'New Order' regime which have influenced the agricultural and rural development strategy adopted in Java and which help to explain the nature of agrarian changes in the recent period of relatively rapid agricultural growth on the basis of an 'unreformed' agrarian structure. This brief general overview does no justice to the many regional and local variations to be found, even in major rice-producing regions, but can serve as backdrop to the two more detailed village-level analyses which follow (Chapters 13 and 14).

Javanese Peasant Society in the Colonial Period

Java has long been thought of as a classic example of a relatively undifferentiated rural society consisting mainly of millions of extremely small, largely family-labor subsistence farms. Although the best-known formulation of this perspective is Geertz's *Agricultural Involution* (1963), the view of rural Java as an egalitarian and stagnating rural society has dominated the ideas of colonial administrators and researchers alike since at least the beginning of the nineteenth century. Raffles, after completing five years as British Lieutenant-Governor of Java (1811–1816), wrote that:

> The lands on Java are so minutely divided among the inhabitants of the villages, that each receives just as much as can maintain his family and employ his individual industry. . . . There is but little inducement to invest capital in agriculture, and much labor must be unprofitably wasted: as property is insecure, there can be no desire of accumulation; as food is easily procured, there can be no necessity for vigorous labor. There exist . . . few examples of great affluence or abject distress among the peasantry; no rich men, and no common beggars.
>
> (Raffles (1978 [1817]): 147)

One of his successors, the Belgian Vicomte Du Bus de Gisignies (Commissioner-General, 1826–1830), shared these views. His lengthy report in 1827 blamed rural stagnation on the system of communal land tenure which was not conducive to economic agricultural growth; the whole population, according to him, consisted of small peasants holding tiny plots barely sufficient to meet their subsistence needs and unwilling to cultivate more lucrative crops than rice for sale on the world market.

Governor-General van den Bosch (1830–1833) held similar views of the Javanese peasantry, although he became Du Bus's main opponent in the heated debate in the Netherlands on how best to make the colony profitable. Unlike his predecessor—who favored economic liberalism, freeing the land and people from their 'communal bondage' and allowing European entrepreneurs to establish corporate plantations—van den Bosch proposed the policy of state exploitation of the colony which came to be known as the Cultivation System, involving forced cultivation of export crops by peasants for delivery at fixed prices and/or against remission of land rent to the colonial government. This system, which van den Bosch was able to institute during his term as Governor-General, continued as the foundation of colonial extractive policy for the next forty years, until the formal opening of Java to private and corporate estate agriculture by the Agrarian Law of 1870.[1]

Colonial policies changed during succeeding years, but prevailing ideas about the social-economic organization of the Javanese countryside have shown a striking continuity. This is perhaps not surprising; comparable attitudes were to be found among most colonial governments in Asia, Africa, and the Americas, as Clive Day noted in his generally positive report on *The Policy and Administration of the Dutch in Java* (1904).

In the late colonial period Boeke's accounts of economic 'dualism' in the oriental economy (Boeke 1927, 1953) can be seen as a twentieth-century mutation of earlier colonial views. Boeke quotes Du Bus at length and agrees with most of his views; the concept of 'static expansion' (increase in numbers at low and stagnating levels of living, without social and economic modernization or differentiation) is a clear echo of Du Bus's account. More remarkable is the persistence and extension of these notions of the village economy since independence. Geertz's *Agricultural Involution* (1963), which has earned much acclaim as well as critique (see, for example, Alexander and Alexander 1982; Kahn 1985; White 1983; but also Geertz 1984), reflects basically similar ideas of the stagnant, subsistence-minded, poverty-sharing and homogeneous 'post-traditional' village society incapable of modernization and economic growth.

Against these views there is ample evidence that social classes based mainly on unequal access to land have existed in Java for a long time. While this evidence is often confusing and difficult to interpret, it suggests that debate should properly concentrate on the nature, implications, and fluctuating dynamic of agrarian differentiation processes at different periods rather than their presence or absence. Archival and published sources for the late eighteenth and early nineteenth centuries provide many examples from different regions pointing to the existence of three broad agrarian classes: first, a substantial group of landless peasants, sometimes attached to landed peasant households but also frequently constituting a more mobile group of itinerant free laborers; second, a large mass of peasants with rights to land and with heavy tributary and corvée obligations attached to those rights, and above them a third class of village officials who in addition to their own holdings had control of a large proportion of village lands as salary lands, plus the rights to the unpaid labor of the landholding *sikep* or *kuli* to cultivate it (Breman 1980, 1983; Carey 1986; Knight 1982). Wage-labor arrangements as well as sharecropping between landless and landholding peasants 'can be said to have been commonplace' (Knight 1982:126; cf. also *Bijdragen* 1862).

The same may be said of agrarian commercialization. Many of the wet-rice regions of Java (and especially of its north coast) were already incorporated into commodity production (mainly sugar, rice, and indigo) for foreign markets since the early eighteenth century. With the introduction of the Cultivation System—which differed more in intensity than in form from previous modes of extraction under the Dutch East India Company—commodity production extended further both in the lowlands (mainly sugarcane) and in the uplands (tree crops, especially coffee, which involved about three times as many households as sugarcane throughout the Cultivation System, cf. Van Niel 1980). Subsequent decades saw the expansion of new commercial crops (soybeans and groundnuts in particular) as secondary crops on peasant rice fields. In the north coast regions, for example, the evidence accumulated by Knight attests to the existence of lively markets in land (both for sale and lease) and peasant produce (including rice), as well as the widespread of 'free' wage labor in both food and export crop production (Knight 1982).

Although there is much debate on the influence of the Cultivation System on patterns of social-economic differentiation, recent studies at the regional level document the consolidation of social inequalities and link these to the enormous increase in the flow of money into village economy in the form of payments for produce and labor, a feature of the system often ignored in earlier literature (Elson 1984:80; Knight 1982:149).

Many observers noted the tendency to concentration of landholdings and wealth and the growth of landlessness during the late nineteenth and early twentieth centuries. Meyer Ranneft, whose observations in many parts of Java were widely quoted at the time, relates these processes to the development of money economy and particularly of moneylending, 'a typical symptom of the forward progress of money-traffic into the peasant household, of the "demonic" power of money' (Meyer Ranneft 1974 [1919]:21). The extent of monetization of the Javanese rural economy became manifest when the Great Depression of the 1930s dealt a crushing blow to commercial agriculture and the various agro-related enterprises (crop processing, transport, and trade in particular) that depended upon it, as we shall see below.

The earliest data collection efforts offering a somewhat systematic picture of the distribution of land and other indicators of rural differentiation date from the beginning of this century. The massive 'Inquiry into the Declining Welfare of the Native Population of Java and Madura,' undertaken in 1904–1905, permits a quantitative approximation of the distribution of land (and of landlessness) as shown in Table 12.1. Despite the inevitable inaccuracies in these estimates and problems in their interpretation (see the notes to Table 12.1), they provide a crude insight into the agrarian structure which attests to the inequality of access to land. Most striking is the high proportion of landless—more than one-third of the agrarian population and more than 40% of the total—while at the other end of the spectrum 9% of owners (5% of the total population) with holdings larger than 2 *bahu* (1.4 ha) owned between them more than one-third of all the land. The qualitative reports of the inquiry mention many local instances of land concentration, and it is often reported that this concentration was greater than ownership figures suggest, since it tended to take the form of other means of gaining effective control of land—leasing, mortgage, and 'purchase with right of repurchase'—rather than outright purchase. The inquiry's reporter from Limbangan (Priangan, West Java) notes, for example: "This evil is grad-

TABLE 12.1. *Estimation of Land Distribution and Landlessness Among Javanese Households from the "Declining Welfare Inquiry' of 1904–1905*

	Number (thousands)	Percent of all owners	Percent of all households
1. Landowners			
Less than 0.5 ha	1395	47	25
0.5–1.4 ha	1318	44	24
1.4–2.8 ha	181	6	3
More than 2.8 ha	90	3	2
Subtotal	2984	100	55
2. Other occupations not combined with own farm enterprise			
Share tenants	267		5
Wage workers in native agriculture	1215		22
Wage workers in foreign-owned estates	167		3
Nonagricultural occupations			
Own account workers	538		10
Hired workers	300		5
Subtotal	2487		45
Total	5471		100

Source: Hasselman (1914:36 and apps. E and L).
Notes: '*Landowners*' in this table comprise the aggregate of two categories: (1) 'individual owners and those with a permanent share in communal land' (totaling 3.39 million) and (2) those with a share in 'periodically redistributed communal land' (0.68 million). The aggregation may result in some overestimate, since the categories may to some extent overlap; for example, an individual owner might also have a share in periodically redistributed communal land. For the '*other occupations*' we have included only males and only those who do not combine these occupations with ownership of farmland, so as to approximate as far as possible the numbers of households and avoid overlap with landowners. These procedures seem justified since the total included in this way (5.47 million) is only slightly larger than the total number of households recorded in the same survey (5.33 million); see Hasselman (1914:app P). The '*nonagricultural occupations*' include trade, land and water transport, handicrafts, and small industries (Hasselman 1914:app. L).

ually increasing. [The statistics on landownership] do not show us the half of the real situation. Widespread pawnings and suchlike arrangements are kept secret by the officials, so that the figures are far too low" (Declining Welfare Inquiry IXa, Preanger Regentschappen, 1907:18). The impression provided by the inquiry, then, is one of appreciable concentration of effective control of land (though with much regional and local variation), through concealed forms of usury and less frequently by outright purchase, into the hands of a well-to-do minority. The identity of this minority is described by nearly all the local reports as some combination of the following: 'village heads and other officials, religious (Islamic) teachers and *hadjis*, and well-to-do villagers'; Chinese, Europeans, and other foreigners are rarely mentioned, although the inquiry included questions specifically about their involvement in these forms of land accumulation.

The corollary of land concentration, as summaries from the inquiry also note, was the increasing proportion of near-landless and landless households. These trends are assessed in quite contradictory ways by the various authors of the inquiry's reports. On the one hand they conflict with the picture of generalized, undifferentiated 'dwarf-farming' given in many other reports of the time and even in some of the Inquiry reports, whose authors seem at times to have ignored their own findings: "For the rest we observe that definitely abundant landholdings are found only among a very small number. As a rule, for the average native one can only distinguish 'adequate' and 'inadequate' holdings (or possibly no holdings at all), 'adequate' here being understood as an area which under normal conditions yields enough to provide the peasant and his household with their daily needs" (Declining Welfare Inquiry IXa, Part I, 1911:71). In other places the reports note the existence of differentiation and large holdings in rural areas and their potential advantages if combined both with agricultural modernization (for the landowners) and improvement of other income sources for those without land (Declining Welfare Inquiry Xa, Part I, 1914:310, 314).

Twenty years later, after rapid expansion of export-crop production and the years of the so-called Ethical Policy with its emphasis on 'education, irrigation, and emigration,' a large-scale inquiry into the 'Economic Condition of the Native Population' concluded that the effect of the boom years had been to increase economic inequalities in rural areas: landless laborers and peasants with insufficient holdings were both worse off in 1924 than they had been in 1913, while the larger landholders' position had improved (Fievez 1926:152). Another inquiry into the 'Tax Burden on the Native Population' from the same years provides a picture of agrarian inequalities not unlike that of the Declining Welfare Inquiry, although this survey used 'welfare' rather than landholdings as its basis of stratification. Its results (summarized in Table 12.2) again suggest that more than one-third of the population were landless, with almost

TABLE 12.2. *Social Stratification of the Native Population of Java and Madura According to the 'Inquiry into the Native Tax Burden' of 1923*

Occupational category	Average annual household income (guilders)	Percentage of households
1. Officials, village heads, religious teachers	n/a	4.0
2. Native wholesale traders and industrialists	1130	0.3
3. Wealthy peasants	1090	2.5
4. Permanent workers in European and Chinese enterprises	370	2.4
5. Middle peasants	300	19.8
6. Retail traders, artisans	248	5.9
7. Poor peasants	147	27.1
8. Sharecroppers having no land of their own	118	3.4
9. Coolies in foreign enterprises	120	19.6
10. Laborers in native agriculture	101	12.4
11. Miscellaneous	n/a	2.6

Source: Adapted from Meyer Ranneft and Huender (1921:10) by Wertheim (1964:112).
Note: The precise criteria for the stratification of 'peasants' into three groups are not clear. From other discussions of this period it is probable that 'poor peasants' are those with less than 0.5 ha or perhaps 0.7 ha (1 *bahu*), 'middle peasants' those with 0.5 (or 0.7) to 1.4 ha, and 'wealthy peasants' those with more than 1.4 ha.

as many 'poor peasants' whose incomes were not greatly higher than those landless laborers and share tenants; at the other end, a smaller number of 'middle peasants, wealthy peasants, native wholesalers and industrialists, officials and village heads, and religious teachers' (about one-fourth of all households) had incomes three to ten times those of landless households (Meyer Ranneft and Huender 1921).

It was this minority of independent ('wealthy' and 'middle') peasants whom Boeke and other administrators saw in the 1920s as the primary potential motor for the development of the 'native' side of Java's economy. Boeke's proposal that colonial welfare policy should be directed not at the whole population but at the group of 'strong, energetic and developed peasants . . . the best among the native population' (Boeke 1927:191) was prompted by the failure of the Ethical Policies; now, the main target for government assistance must be those who had sufficient resources and skills to make optimal use of modern agricultural extension services and new production techniques—a principle that would reappear forty years later in the practice, if not in the stated policies, of the 'New Order' state.

This group of 'progressive peasants' had made great advances through their increasing entrepreneurial activities during the period of relatively rapid economic growth after the First World War, through increasing agrarian commercialization and the strengthening of village elites whose interests were fostered by the colonial policies of the time. Geertz notes in his social history of 'Mojokuto' (East Java) that rapid expansion of the sugarcane economy and other agrarian developments accompanying it had by the 1920s stimulated the concentration of holdings on the one hand and the proletarianization of marginal peasants on the other; the larger landholders (including in Mojokuto several with holdings of about 50 ha) involved themselves both in 'rentier' and 'entrepreneurial' ventures, as labor hirers, harvest contractors, and intermediaries in the acquisition of rented peasant land for the sugar factories, as lenders of money or consumer goods at exorbitant interest rates, as purchasers of agricultural equipment and draft animals for resale, loan, or hire to small peasants or their own tenants, and as first-level traders in the growing volume of dry-land peasant cash crops destined eventually for Chinese wholesalers (Geertz 1965:39ff.). This period also saw an incipient rationalization of farm management in paddy cultivation through labor cost-saving practices remarkably similar to those reported later in the 1970s: lower in-kind (*bawon*) wages for harvesters and the introduction of closed labor markets through exclusionary recruitment practices such as *tebasan* and *kedokan* (Burger 1930:33f.; van der Kolff 1937:17f.).[2]

These processes came to an abrupt halt during the depression years of the 1930s, which caused a drastic reduction in wage employment in both plantation and peasant agriculture, with trade in agricultural products also declining or giving way to barter in the face of general demonetization.[3] The years of distress were all the more marked as they had followed a period of relatively rapid commercialization, and the more intensive commercial involvement of village elites was in turn the main reason for the great reversals suffered by this group during the depression (Elson 1984:242ff.). In a study of several 'remote' villages in East Java, van der Kolff reported that by 1922 rich peasants had come to dominate the rural economy and that wage labor was replacing traditional forms of reciprocal labor and mutual aid; revisiting the same villages in the mid-1930s he found that social differentiation had decreased and a system of 'shared poverty' had emerged, as the rich lost most of their capital as well as their control over labor (van der Kolff 1937).

The agrarian decommercialization occasioned by the depression had scarcely begun to reverse itself when it was further reinforced by the renewed collapse of the export crop industries and demonetization during the Japanese occupation and ensuing national revolution (1942–1949). In the immediate postindependence years, rural commercial development began again for a short time, although within a severely constrained macroeconomic context. Various studies from this period provide evidence of marked inequalities in access to land and a rather pronounced social differentiation (cf. the summaries of various studies provided in Jaspan 1961 and Lyon 1970 and the Indonesian Communist Party's studies in various parts of Java in the late 1950s, also summarized in Lyon 1970:20ff.). In the subsequent years of 'Guided Democracy' and 'Guided Economy' (1957–1965) the decline of export-crop economy, the problems of hyperinflation, and the political turmoil over land reform implementation were in turn powerful countervailing forces, although it is by no means clear that the relative position of wealthy villagers deteriorated in this period:

> With the deterioration of rural economic conditions—the increase in rural indebtedness, the consequent demand for cash and, more broadly, the . . . changes in land distribution—the roles of richer peasants and large landlords have undergone a shift. This shift has been toward their greater relative financial advantage, so that in their functions as moneylenders, hirers of wage labor, purchasers of crops and so forth, landlords are operating for the most part under very favorable bargaining conditions.
>
> (Lyon 1970:26)

Nevertheless it is only after the coming to power of the present 'New Order' regime and its comprehensive program of political and economic reconstruction, backed with substantial external funds, that we can speak of agricultural policies and associated structural changes in a context of overall economic growth linking up with processes set in motion in the 1920s (Hüsken 1984).

The long history of inequalities in land control, wealth, and power in Javanese society which we have summarized in the preceding paragraphs leaves us still with only a vague picture of the concrete forms of social-economic differentiation; we need particularly to know whether land concentration was accompanied by tendencies to expansion in the scale of peasant enterprises and the emergence of capitalistic wage-labor farming, or rather by a proliferation of small tenancies dependent on a rent-capitalistic gentry not themselves involved in the management of farm enterprises (the latter being the path which has recently been suggested as historically typical of Asian wet-rice societies; see Bray 1983). Discussion of this question is severely hampered by the fact that colonial data (such as those shown in Table 12.1), while providing a picture of landownership patterns, do not tell us about trends in operated farm size and its distribution.

In a 1936 dissertation heavily influenced by Chayanov, Ploegsma reviewed a large number of local-level studies and reports and concluded that recent trends towards concentration of effective control of land in Java were not in general accompanied by the emergence of large, capitalistic farm enterprises but rather by increasing tenancy and the maintenance of the pattern of small, subsistence-oriented, Chayanovian 'family-labor farms.' The same was true, he argued, for other rural enterprises such as crafts, industry, and trade (Ploegsma 1936:186).

This argument, however—like many others on both sides, before and since—stretches available information too far in order to arrive at a simple characterization of an agrarian situation that was in fact more complex. On the one hand, it appears true that very large owners (with holdings running into hundreds and in a few cases thousands of hectares) generally did not run large capitalist wage-labor farms, as for example the unusually large landowners in the Priangan region of West Java, whose holdings had been acquired through usurious practices since the late nineteenth century (Ploegsma 1936:61). Other studies, however, including those used by Ploegsma, indicate that alongside 'feudal-aristocratic' landlordism large farm enterprises also existed (Scheltema, Vink, Pertjaja, and Gadroen, cited in Ploegsma 1936:63). Additional evidence of the existence of larger, wage-labor based enterprises alongside millions of small owner- or tenant-operated holdings may be seen in the relatively large numbers of agricultural wage-labor households working in the peasant sector—for example, in the 1923 inquiry summarized in Table 12.2, landless 'laborers in native agriculture' number about one-fifth the number of 'peasants.' We may also mention the numerous village-level studies by Vink and his colleagues in the 1920's which found hired labor averaging about 60% of all labor inputs in peasant rice cultivation (summarized in Collier 1981b).

Agrarian relations in the late colonial period, then, cannot simply be characterized either in terms of oppositions between a Geertzian 'effete gentry' (Geertz 1965:49) and a mass of small owner-peasants and tenants from whom they extracted rent or interest or in terms of an agrarian bourgeoisie operating large commercial farms by the extraction of surplus value from an agricultural proletariat. Both kinds of relationship were clearly present, but firm conclusions about the relative importance of each in different regions and at different periods are not possible until more detailed historical work is done.[4] It is clear, though, that more recent social differentiation in the countryside and the growth of a commercial type of peasant agriculture are neither the result of abrupt technological changes transforming a previously homogeneous peasantry nor the product of progressive linear tendencies of continuous penetration of capitalism into village life. Instead, commercialization and 'de-commercialization' and the changing agrarian relations which accompanied them seem to have been cyclically alternating responses to changing conditions of the outside market which determined their course and pace.

Political Control and Production Strategies: 1966–Present

During the years 1880–1980 Java's population grew from about 18 million to about 90 million; this fivefold increase reflects continuous but not extraordinarily high population growth rates fluctuating between 1% and 2.5% per year. During this period paddy and other food crop production also grew rather steadily but failed (until the mid-1960s) to keep up with population growth, the overall result being a long-term decline in per capita paddy output from about 0.20 ton in 1880 to an all-time low of 0.13 ton in 1965 (Booth 1985). Paddy yields on irrigated land were also remarkably stable at around 2.1–2.2 tons per hectare, reflecting a basically unchanging but already extremely labor-intensive technology. Paddy production growth during the past 100 years can be broadly divided into three periods. Before 1900, almost all production growth was achieved simply by expansion of the area of irrigated land; between 1900 and 1960, area expansion was accompanied by increases in double-cropping, with the latter gradually assuming more importance. It is only since the 1960s that yield growth through the large-scale incorporation of new inputs has come to play the main role in increasing production (Booth 1985:36).

In trying to characterize agricultural growth strategies during the past twenty years and the agrarian changes that have accompanied them, a number of factors must be taken into account. First are the already sharp inequalities in access to land, wealth, and power which were the basis of the widespread agrarian conflicts during the relatively democratic rural political atmosphere of the Sukarno era. Access to land and agrarian relations were the focus of intense political struggles in many regions of Java in the early 1960s, in which the Indonesian Communist Party (PKI) and its peasant, women's, and youth affiliates (BTI, Gerwani, and Pemuda Rakyat) played the leading offensive role. These conflicts centered mainly around the struggle for implementation of Indonesia's rather moderate Agrarian Laws of 1960 and Share-tenancy Regulations of 1959, although in some areas the BTI initiated a more radical and threatening 'land to the tiller' campaign alongside unilateral actions to enforce the reform implementation, bypassing the ineffective local Land Reform Committees. There is also some local evidence of actions to ban certain forms of injurious labor-tying arrangements such as *kedokan/ceblokan* (see note 2) and to maintain levels of in-kind harvest wages, issues which were of particular importance to landless and near-landless women. These campaigns were the main cause of the extraordinarily violent backlash in 1965–1966 by Moslem organizations and larger landowning groups, preceded, assisted, or provoked by the military in which hundreds of thousands of unarmed PKI members or suspected sympathizers were massacred and others detained without trial for many years (Mortimer 1974).[5] One consequence of this bloody and tragic watershed was that Java's Green Revolution would later be implemented on the basis of an 'unreformed' agrarian structure.

Secondly, we must bear in mind Indonesia's long history of chronic food production deficits. Since independence, rice imports (destined mainly for the cities) had risen from about 0.3 to about 1.0 million tons (or about 10% of domestic consumption) in the early 1960s, dropping drastically to 0.2 million tons in the crisis year 1965 (Mears 1981:27). Severe food scarcities caused by the combination of production shortfalls and the incapacity of a collapsing economy to afford sufficient imports, along with the hyperinflation of the late Sukarno years, were in the last analysis more important causes of the collapse of the Sukarno regime than the so-called 30 September Movement, which in more favorable economic circumstances might have proved a crisis no less manageable than others which Sukarno had overcome during twenty years as President. Rapidly rising urban food prices are closely linked to urban political unrest in Indonesia as in many other countries. The agricultural and rural development strategies of the 'New Order' regime should then be viewed in the context of two overriding imperatives: first, the need to achieve and maintain political control in rural areas and, second, the need to ensure food availability at relatively stable prices in the cities. Both of these imperatives were (and still are) necessary conditions for the survival of the 'New Order' regime and in turn for the continued flow of foreign aid and investment and the continued access of those in power to the surpluses generated in the reorganized production system through the extraction of various forms of monopoly rent.[6]

It is not surprising, then, that agricultural and rural development strategies under the 'New Order' regime provide an almost classic example of the 'top-down' approach (cf. Hardjono 1983). The total destruction in 1965–1966 of the Communist Party and its rural mass organizations—which, whatever their strategic errors (Tornquist 1984), had been 'the most important agent of change threatening the duality of poverty and privilege' (Palmer 1978:6)—was followed by a long process of reorganization of rural political and institutional life. We may first mention the doctrine of the 'floating mass,' a term coined in 1971 to denote the effective cutting off of the rural masses from any organized political-party involvements and formally enacted in a law of 1975 banning all political party branches below the level of the district capital town. The Agrarian Reform laws of 1960, although formally remaining on the books, have been tacitly abandoned, and any efforts to protest against illegal attempts to expropriate land have in general been ruthlessly suppressed. The peasant and agricultural-worker affiliates of all former political parties have been replaced by the monolithic state-sponsored 'Indonesian Farmers' Unity Association' (HKTI), and all other interest-based associations (of fishermen, women, youth, etc.) have been replaced by similar state-sponsored substitutes. Such diverse forms of rural cooperative associations as previously existed (which again were frequently linked to political parties in the Sukarno years) have been channeled into a single approved form, the 'Village Unit Cooperative' (KUD); the formation of independent cooperatives for any rural economic activity, whether agricultural or not, is formally disallowed by Presidential Instructions of 1978 and 1984.

Decision-making institutions within the village itself have lost much of the democratic and participatory character of the Sukarno years. Higher-level state officials often interfere with procedures for the selection of village officials, sometimes by imposing military personnel without election and sometimes by screening out popular candidates where elections are still held. Other village-level decision-making institutions have been dismantled and revised, culminating in the new Regulations on Village Government (UUPD) of 1979: popularly elected village representative councils have been replaced by the 'Village Social Security Institute' (LKMD), whose members are appointed by the village head in consultation with the subdistrict head (Schulte Nordholt 1986). These and other changes have greatly increased the power of the village heads—or more accurately restored them to a level of power which they had enjoyed in colonial times but had been much eroded by the more participatory institutions and atmosphere of local politics and administration after independence.

The impact of these changes on the character of rural life is well encapsulated in a study of a village near Yogyakarta by the Indonesian rural sociologist Loekman Soetrisno:

> The village administration are once again dependent on the district head as they are accountable to him rather than to the peasants. . . . The abolition of the political party at the village level has deprived the peasantry of the 'protective umbrella' from state officials and other abuses. . . . It also deprives them of their rights to organize and collectively pursue their own interests and strengthen their bargaining power vis-à-vis the local village elite. . . . Asked to evaluate these changes, peasants in Selomartani say they miss the unity between peasants and village administration that they saw prevailing 15 years ago. Then, they say the *lurah* (village head) and his assistants were more concerned [with] the welfare of the peasants and they showed more respect to them. Nowadays, they say further, the *lurah* and his assistants are more concerned with collecting taxes and ordering the peasants around to join the various development programs that are introduced by the central government.
>
> (Soetrisno 1981:9f.)

Such impositions of monolithic, 'top-down' forms of rural organization are of course not uncommon in the recent political history of the ASEAN region or in other parts of Asia, Africa, or Latin America. What is more remarkable is that the absence of any political institutions or mass organizations linked with peasant interests—there is nothing, for example, comparable to the powerful farmers' lobbies of India—has not in this case been accompanied by neglect of small-scale agriculture in state spending, although it has determined the 'top-down' way in which programs directed at peasant producers are implemented.

The other element in the picture, which has enabled state imperatives to be realized, is that the 'New Order' regime has not only recognized the importance of food availability and prices for its own survival but has also commanded the resources to make this possible on an unprecedented scale. International grants and loans channeled through the 'annual IGGI fix'[7] have provided, each year since 1968, amounts far greater than the total state income or expenditure in any year of the Sukarno era; to this we must add the even greater sums available since the OPEC windfall of 1974 which caused the price of Indonesian crude oil, the source of about 70% of state revenues, to rise from US$3 to $12 a barrel within one year (and later by 1982 to US$36). These revenues have made possible levels of development spending which in recent years amounted to about US$300 per year for every Indonesian household. During the 1970s, about 20% of the development budget was allocated to agricultural development (including a total of almost US$2 billion from the IGGI donors up to 1984); equally impressive amounts were channeled to rural areas in the form of infrastructure development programs (Cooke 1986). It is mainly these resources which made possible the eventual success of Indonesia's (and Java's) Green Revolution in rice production, which had got off to an extremely shaky start many years earlier.

During the Sukarno era we may identify a 'proto–Green Revolution' phase (Franke 1972) in which various attempts were made to provide credit and extension to rice farmers and to stimulate the use of artificial fertilizers and 'National' (i.e., pre-IRRI) improved seed varieties. In the later Sukarno years students from various universities, working with minimal material support, initiated and expanded a 'mass guidance' campaign among peasants which seems to have been successful in many regions; average paddy yields in Java for the first time surpassed those of the colonial period (averaging 2.4 tons/ha in 1961–1965) with more spectacular results in the concentration areas of the intensification campaign. In the immediate postcoup years the new regime paid little attention to agricultural production problems, but after a series of poor harvests and a severe shortage in 1967 (the main cause of the rapid inflation of 40% in January 1968 which threatened the economic stabilization program) and accompanying urban riots a series of efforts to increase food supplies were initiated, both by increasing imports of rice and wheat and by attempts to increase domestic rice production. In a new-style rice intensification program known as '*Bimas Gotong Royong*' (*Bimas* = mass guidance, *Gotong Royong* = mutual cooperation), the state contracted with a number of multinational corporations (paying them about US$50 per hectare) for provision of the necessary Green Revolution inputs including fertilizers, insecticides, extension and management, and the new IR varieties which had recently become available from Los Banos; peasants were expected to repay these inputs by delivering one-sixth of their crop to a national collecting agency. By all accounts mismanagement and corruption made a near-disaster of this scheme, which was abruptly terminated in 1970 when the state took direct control of the *Bimas* and *Inmas* (mass intensification) programs, which continued with various modifications up to the early 1980s (Franke 1972; Hansen 1971; Rieffel 1969; Roekasah and Penny 1967).

It was not, however, until after another rice crisis (when poor harvests in 1972–1973 coincided with sharply rising prices in the world market and rice imports more than doubled from 0.74 to 1.66 million tons in 1973) that we can see the state shifting from a 'command' production strategy to a more effective but also much more costly one of 'command plus subsidization,' now made possible by rapidly rising state revenues. The main elements in the subsidies (described in detail in Mears, 1981) have been:

1. Massive subsidies on artificial fertilizer prices, which increased the fertilizer/rice-price ratio (kg of urea that can be bought with 1 kg of paddy) from 0.6 in 1974 to 1.9 in 1982 (Dapice 1985). This subsidy was costing around US$0.5 billion annually by the early 1980s.
2. Subsidized agricultural credit through the *Bimas* and *Inmas* programs. The 'subsidy' has in practice included not only interest rates near or below inflation, but also the tolerance of high levels of nonrepayment, with larger farmers, village officials, and 'fictive farmers' (those who through local connections obtained credit for farms which did not exist) being the outstanding defaulters. Credit distribution reached its peak in the mid-1970s (although still reaching only a minority of larger rice farms) and in 1983 collapsed in a mass of unpaid debts.

3. State purchases of paddy through the operation of a floor-price intervention scheme and through purchases at market rates when prices are above official floors in order to build up and maintain national buffer stocks (now at about 3 million tons). These efforts, which have also been costly, have contributed to favorable real rice prices to Javanese rice producers, with real rice prices increasing by about 50% during the 1970s.
4. Increasing quantities of free or subsidized irrigation water provided by the many irrigation rehabilitation and development schemes financed by foreign borrowings.

The results of these subsidies in terms of production are not surprising. Javanese peasants now produce more than twice as much rice as they did in the late 1960s from a harvested area which has risen only marginally; Java has made more than an average contribution in terms of yield growth compared to other regions of Indonesia and has therefore played a major role in Indonesia's transition from the status of the world's largest rice importer to a situation (1985) in which self-sufficiency and even a modest surplus can now be expected in normal years.

The political advantages of this costly rice-production drive go beyond the elimination of import dependency and the power to control urban food prices. The inequalities of Java's 'unreformed' agrarian structure (as we will see in Table 12.4) mean that some 3 million rural households (one-third of all farm operators, one-fifth of all rural households) farm almost four-fifths of all irrigated riceland; even allowing for the modest inverse relationship between farm size and yields (cf. Keuning 1984:61) about three-quarters of all rice produced in Java is grown on their farms. These larger farmers, who have also been the main recipients of *Bimas* credit and other subsidies, have therefore been the main beneficiaries of the Green Revolution and of the subsidies we have described. State patronage of this group of larger landholders, it has been argued (cf. Hart, Chapter 2 and Hüsken, Chapter 14), has in turn eased the problem of political control in rural areas, which even a large military and police presence cannot accomplish without civilian allies in a rural population of some 70 million.

After the first years of 'New Order' government, when the short-lived alliance of armed forces, pious rural Moslems, and urban intellectuals who had combined to oust Sukarno and crush the Communist Party had been dissipated, the latter groups found themselves basically as excluded as any other civilian groups from the political process, and the loyalty of rural elites could no longer be assumed. Rural elites, however, have more or less consistently maintained their loyalty and participated in the 'self-policing' of rural areas for a number of related reasons. First, of course, because of the direct benefits accruing to them in the form of subsidization of their farm enterprises, but also because of the advantages of a political context which stifles all forms of organized struggle against continuing or increasing agrarian inequalities. This context has, for example, permitted a large number of rationalizations in labor process in the peasant sector to be accomplished—if not without local instances of resistance, at least without major political struggle.

These rationalizations have included the widespread introduction of laborsaving technologies such as replacement of the finger-knife (*ani-ani*) by the sickle in harvesting; rotary or toothed weeders in place of hand weeding; diesel-powered rice hullers in place of hand pounding, and in some areas threshing machines in place of hand flails and tractors in place of the hand hoe or animal-drawn plow. Besides these technical changes, shifts have been widely reported in hired labor recruitment and payment practices which have functioned to replace traditional 'open' labor markets with exclusionary practices such as *tebasan* and *kedokan* (see note 2) in the most labor-intensive operations of ground preparation, transplanting, weeding, and harvesting (Collier et. al. 1974; Collier 1981b; Kanō 1980; Hayami and Kikuchi 1981). Both men and women have been affected by these changes, but their impact on women in landless and near-landless households has perhaps been greater due to their great reliance on wage incomes in transplanting, harvesting, and postharvest work (Pudjiwati Sajogyo 1985; White 1985).

These changes have had considerable impact on the distribution of agricultural incomes in the 'unreformed' agrarian structure, in which a minority control the greater part of farmland but about 80% of labor inputs in rice cultivation in the late 1960s consisted of hired labor provided by the majority of landless and near-landless households (White 1985; cf. Collier 1981b). Both large-scale data (Papanek, 1985) and village-level surveys (White and Wiradi, Chapter 13) have found that during this period of relatively rapid production growth the share of output received by hired laborers in the form of wages has declined proportionately to the much more rapid growth of the farmers' net income from crop production.

If colonial inquiries frustrate us by providing data on landownership but not on farm size distribution (see above), the problem since independence is reversed. Three successive decennial Agricultural Censuses provide data on farm size distribution, but there are no comparable large-scale data on landownership. In Table 12.3 we have summarized data from 1963, 1973, and 1983 on the distribution of operated farm sizes (excluding holdings of less than 0.1 ha, which as Table 12.4 indicates comprised more than 1 million households but only 1% of farmland in 1983). This comparison over twenty years gives an impression of almost uncanny stability, with shifts of only a percent or two in any group and the same basic pattern persisting with about half of all land being farmed by about 20% of farm households with holdings of 1.0 ha or more. These data in fact serve mainly to highlight the limitations of farm-size statistics in telling us anything about agrarian change, especially in contexts of population and productivity growth (cf. White, Chapter 1).

First we should remember that these data cover only households with access (whether as owner-operator or tenant) to a 'farm.' While their number has grown by about 16% over twenty years, total rural population was growing much faster at about 2% per year. Extrapolating from available population censuses covering roughly the same period (1961, 1971, and 1980), it appears that about 73% of rural households had a farm (of more than 0.1 ha) in 1963 and only about 57% in 1983, suggesting a rather dramatic increase in absolute landlessness which village-level studies also confirm (White and Wiradi, Chapter 13). Secondly, if we consider the larger farms whose numbers have grown absolutely but not proportionately during twenty years (for example, those with more than 1.0 ha who farm more than half of all the land in the peasant sector), the apparent stability of average farm sizes among this group masks an enormous increase in the capacity of these farms to provide those who control them with an agrarian surplus. Between 1963 and 1983 average yields in paddy cultivation in Java rose from about 1.3 to 3.1 tons of milled rice per hectare (Mears 1981, 1984); a holding of stable size now represents a quite different proposition in terms of the household's income, particularly in a context of subsidized production. When increased cropping ratios are combined with yield increases, the difference is even more striking, as vividly illustrated by a recent restudy in a village in East Java. For the 22% of households in this village which still owned irrigated rice fields in 1981, a half-hectare of *sawah* which produced a total of 3.6 tons of paddy per year in 1965

TABLE 12.3. *Distribution of Operated Farm Sizes: Java, 1963–1983*

	1963			1973			1983		
	% of farms	*% of area*	*Average size (ha)*	*% of farms*	*% of area*	*Averege size (ha)*	*% of farms*	*% of area*	*Average size (ha)*
Operated farm size (ha)									
0.1–0.49	52	25	0.3	55	22	0.3	54	21	0.3
0.5–0.99	27	27	0.7	26	27	0.7	27	27	0.7
1.0–1.99	15	23	1.1	14	27	1.3	14	27	1.3
2.0 and above	5	25	3.3	5	24	3.2	5	25	3.1

	1963	1973	1983
Total farms ⩾ 0.1 ha (millions)	7.95	8.66	9.21
Total area (million ha)	5.65	5.51	6.17
Average farm size (⩾ 0.1 ha)	0.71	0.64	0.67

Source: Central Bureau of Statistics, Agricultural Census (1963, 1973, 1983).

Note: To allow comparison, definitions have been standardized between the three censuses and farms of less than 0.1 ha (which were not recorded in 1963 and 1973) have been excluded. This accounts for some differences between 1983 data in this table and Table 12.4. Percentage totals do not always equal 100 due to rounding.

TABLE 12.4. *Distribution of Land Controlled Among Rural Households: Java, 1983*

Area controlled (ha)	No. of rural households (millions)	%	% of all land	Average area (ha)	No. operating *sawah* (millions)	% of all *sawah*	Average area *sawah* (ha)
None	5.52	35	0	0	0	0	0
Less than 0.1	1.04	7	1	0.05	0.40	1	0.05
0.1–0.49	4.98	32	21	0.26	3.53	23	0.19
0.5–0.99	2.45	16	26	0.67	1.89	28	0.42
1.0–1.99	1.31	8	27	1.29	0.92	26	0.81
2.0 and above	0.49	3	25	3.11	0.49	23	1.36
Total	15.79	100	100	0.61	7.24	100	0.40

Source: Central Bureau of Statistics, Agricultural Census (1983).

Note: Area controlled = (area owned) + (area originating from other parties) − (area held by other parties). *Sawah* = wet rice terraces. 'Average area *sawah*' in last column is for *sawah* operators only.

produced 9.6 tons in 1981, 'enough to feed six families at subsistence level' (Edmundson and Edmundson 1983:47-51).

Conclusions

For a number of reasons the agrarian class structure of Java is difficult to categorize in any simple way. Statistics such as those we have discussed attest to the overall small scale of farm sizes, so that virtually all of them would be swallowed up in the 'small-farm' category of any other less densely populated country. However, again and again we are confronted with large-scale or village-level data pointing to the domination of peasant agriculture by a small group of 10–20% of rural households who between them control some 70–80% of all farmland, in holdings which—however small by absolute standards—provide a substantial surplus for those who control them. While 'small peasants' predominate numerically, most peasant-sector *production* is realized on these larger farms and carried out on the basis of wage-labor arrangements between larger farmers and landless or near-landless men and women workers. These larger landholders dominate village-level power structures and have access to state patronage both in and outside agriculture; the combination of agrarian surplus and power enables them also to dominate the growing nonagricultural sector, which recent large-scale surveys suggest provides about half of all rural incomes (even more in some of the villages described in White and Wiradi, Chapter 13). The accumulation strategies of landed elites in this context, then, may involve not the further concentration of landed property but a combination of agricultural intensification and diversification into nonagricultural enterprise on the basis of relatively limited landownership (something not uncommon to 'kulak' classes generally).

At the other end, despite the prevalence of wage labor on the larger farms and the relatively high rates of landlessness, we find no such thing as a pure agricultural proletariat. If diversification is an accumulation strategy for rural elites, for the landless and those with 'sublivelihood' holdings it is a survival strategy to supplement inadequate own-farm or farm-wage incomes. In addition to diversification within the village economy, there has been a rapid increase in temporary (seasonal or more commonly year-round circulatory) migration of household members to urban centers (Hugo 1985), further increasing the range and complexity of the economic circuits and work relations in which individuals, and even more so households, are involved. Agricultural wages and subli-velihood farm production assume proportionately less importance in the incomes of these various kinds of semiproletarian rural households.

The one group that is perhaps hardest to recognize in village society is the small but independent peasantry, producing at a low level for their own subsistence, with a low level of commercialization, neither buying nor selling labor power, neither exploiting nor exploited. Sukarno's classless *marhaen* (Sukarno 1970), who also has epitomized conventional views of Javanese society from Du Bus to Geertz, exists only as a minority and does not in any way constitute the basis of the system of agrarian production.

As we write (1986), Javanese villages are yet again feeling the chill wind of the latest in the long series of economic recessions that have visited Indonesia since the 1880s. With the collapse in oil prices, falling world commodity prices, the international trend towards protectionism, and a deep slump in the industrial sector, by late 1985 it appeared that 'the economy has now clearly run out of steam' (Dick 1985:1). Development spending has been severely curtailed, the urban boom (which provided employment not only for urban workers) is at a standstill, agricultural subsidies are being reduced, and it is likely that real paddy prices will be allowed to fall. Given what we have argued about the political importance of state patronage, it is intriguing, but perhaps not yet productive, to speculate on the likely effects of this retrenchment on patterns of social differentiation and political loyalties in a divided rural society.

Notes

1. In practice there was never a single clearly defined Cultivation 'System' but great variation in its application both between regions and over time (Ricklefs 1981:ch. 11; van Niel 1964). In some regions at least, the transition from forced to 'free' labor was largely accomplished before the regulations of 1870 which formally ended the 'System' (Elson 1984:ch. 4; Knight 1982).

2. *Tebasan*: the sale of standing crops in the field shortly or immediately before harvest to an intermediary (*penebas*) who brings his or her own team of harvesters. These harvesters are often paid with a cash wage rather than the traditional harvest share (*bawon*).

Kedokan (often known as *ceblokan* in West Java and also by a variety of other local terms): a labor-tying arrangement involving the requirement to perform some unpaid preharvest labor (in ground preparation, transplanting, weeding, etc.) in order to have the right to a place in the harvest and a *bawon* wage.

3. For a general overview of the effects of the Depression on rural Java see the report of the Coolie Budget Commission of 1939–1940 (1956); for a regional study, see O'Malley (1977).

4. Compared to the many excellent studies focusing on the nineteenth century, there is remarkably little recent work on the agrarian history of the late colonial period; one such work is Breman (1983).

5. The agrarian conflicts of 1960–1965 and the subsequent massacres in many areas of East and Central Java and the north-coast regions of West Java (described in Mortimer 1972 and 1974) are a vivid reflection of the deep vertical cleavages in rural society. In this light, we find it hard to accept Kahn's recent statement—which he regards as 'self evident'—that 'the political mobilization of Indonesia's rural masses . . . has rarely taken place on the basis of a denial of the homogeneity of rural producers, and has usually been based on the perception of inequalities between peasants and classes defined outside the village economy' (Kahn 1985:79).

6. A useful description of the workings of the 'New Order Fiscal System' is to be found in Race (1981). This model, however, may lately be losing its validity as the private appropriation of public office by military and civilian officials comes increasingly in contradiction with the imperative of industrial growth (Robison 1986a, 1986b).

7. IGGI, the Inter-Governmental Group on Indonesia, comprises the major bilateral and multilateral donor agencies and has met each year since 1967 under the chairmanship of the Netherlands to decide on an aid package for Indonesia. The package currently (1986) amounts to some US$2.5 billion per year.

References

Alexander, J., and P. Alexander (1982). "Shared Poverty as Ideology: Agrarian Relationships in Colonial Java." *Man* 17:597–619.

"Bijdragen tot de Kennis der Inlandsche Huishouding op Java" [Contributions to the knowledge of native economy in Java] (1862). *Tijdschrift voor Nederlandsch-Indië* 24:73–81, 137–143.

Boeke, J. (1927). "Het zakelijke en het persoonlijke element in de koloniale welvaartspolitiek" [Objective and personal elements in colonial welfare policy]. *Koloniale Studiën* 11:157–192. [Translation in W. F. Wertheim et al., eds., *Indonesian Economics: The Concept of Dualism in Theory and Policy.* The Hague: Van Hoeve, 1960.]

——— (1953). *Economics and Economic Policy of Dual Societies as Exemplified by Indonesia.* Haarlem: Tjeenk Willink.

Booth, A. (1985). "Accommodating a Growing Population in Javanese Agriculture." *Bulletin of Indonesian Economic Studies* 21:115–145.

Bray, F. (1983). "Patterns of Evolution in Rice-Growing Societies." *Journal of Peasant Studies* 11:3–33.

Breman, J. (1980). "The Village on Java and the Early Colonial State." CASP Series no. 1. Rotterdam: Erasmus University.

——— (1983). *Control of Land and Labour in Colonial Java.* Verhandelingen van het Koninklijk Instituut voor Taal-, Land- en Volkenkunde, no. 101. Dordrecht: Foris.

Bryant, N. (1973). "Population Pressure and Agricultural Resources in Central Java: The Dynamics of Change." Unpublished Ph.D. dissertation, University of Michigan.

Burger, D. (1930). *Vergelijking van den Economische Toestand der Districten Tajoe en Djakenan* [Comparison of economic conditions in Tayu and Jakenan]. Weltevreden: Kolff.

Carey, P. (1986). "Waiting for the 'Just King': The Agrarian World of South Central Java from Giyanti (1755) to the Java War (1825–30)." *Modern Asian Studies* 20:59–137.

Collier, W. (1981a). "Agricultural Evolution on Java." In G. Hansen, ed., *Agricultural and Rural Development in Indonesia.* Boulder: Westview Press.

——— (1981b). *Declining Labour Absorption (1878 to 1980) in Javanese Rice Production.* Rural Dynamics Study, Occasional Papers no. 2. Bogor: Agro Economic Survey.

Collier, W., Soentoro, Gunawan Wiradi, and Makali (1974). "Agricultural Technology and Institutional Change in Java." *Food Research Institute Studies* 13:169–194.

Cooke, K. (1986). "Big Success in Farming." *Financial Times Indonesia Survey,* 10 March 1986, p. xii.

Coolie Budget Commission (1956). *Living Conditions of Plantation Workers and Peasants on Java in 1939–1940.* Modern Indonesia Project, Translation Series. Ithaca: Cornell University.

Dapice, D. (1985). "Indonesian Food Policy Since 1967: Surprising Progress." Paper presented at SSRC Conference on National Policies towards the Agrarian Sector, Thailand, January 1985.

Day, C. (1904). *The Policy and Administration of the Dutch in Java.* New York: Macmillan.

Declining Welfare Inquiry (1907–1911). "Onderzoek naar de Mindere Welvaart der Inlandsche Bevolking op Java en Madoera" [Inquiry into the declining welfare of the native population of Java and Madura]. *Economie van de Desa,* vol. IXa, part I (Overview), part II (Appendices). Batavia: Kolff, 1911. *Eindverhandeling,* vol. Xa, part I. Batavia: Ruygrok, 1914.

Dick, H. (1985). "Survey of Recent Developments." *Bulletin of Indonesian Economic Studies* 21:1–29.

Du Bus de Gisignies, L. (1827). "Rapport van den Commissaris-Generaal Du Bus over het Stelsel van Kolonisatie, 1827" [Report of Commissioner General Du Bus on the Colonisation System]. In D. Steijn Parvé, *Het Koloniale Monopolie-Stelsel Getoetst aan Geschiedenis en Staathuishoudkunde.* Zaltbommel: Loman.

Edmundson, W., and S. Edmundson (1983). "A Decade of Village Development in East Java." *Bulletin of Indonesian Economic Studies* 19:46–59.

Elson, R. (1984). *Javanese Peasants and the Colonial Sugar Industry: Impact and Change in an East Java Residency, 1830–1940.* Singapore: Oxford University Press.

Fievez de Malines van Ginkel, H. (1926). *Verslag van den Economische Toestand der Inlandsche Bevolking, 1924* [Report on the economic condition of the native population, 1924]. Weltevreden: Kolff.

Franke, R. (1972). "The Green Revolution in a Javanese Village." Unpublished Ph.D. dissertation, Harvard University.

Geertz, C. (1963). *Agricultural Involution.* Berkeley: University of California Press.

——— (1965). *The Social History of an Indonesian Town.* Cambridge: MIT Press.

——— (1984). "Culture and Social Change: The Indonesian Case." *Man* 19:511–532.

Gordon, A. (1978). "Some Problems of Analysing Class Relations in Indonesia." *Journal of Contemporary Asia* 8:210–218.

Hansen, G. (1971). "Episodes in Rural Modernisation: Problems in the Bimas Program." *Indonesia* 11:63–81.

Hardjono, J. (1983). "Rural Development in Indonesia: The Top-Down Approach." In D. Lea and D. Chaudhri, eds., *Rural Development and the State.* London: Methuen.

Hasselman, C. (1914). *Algemeen Overzicht van de Uitkomsten van het Welvaart Onderzoek, Gehouden op Java en Madoera in 1904–1905* [General overview of the results of the Welfare Inquiry held in Java and Madura in 1904–1905]. The Hague: Nijhoff.

Hayami, Y., and M. Kikuchi (1981). *Asian Village Economy at the Crossroads.* Tokyo: University of Tokyo Press.

Hugo, G. (1985). "Structural Change and Labour Mobility in Rural Java." In G. Standing, ed., *Labour Circulation and the Labour Process.* London: Croom Helm/ILO.

Hüsken, F. (1984). "Indonesia's Farm Policy: No Structural Change." *State and Society* 5:73–81.

Jaspan, M. (1961). *Social Stratification and Social Mobility in Indonesia: A Trend Report and Annotated Bibliography.* Jakarta: Gunung Agung.

Kahn, J. (1985). "Indonesia After the Demise of Involution: Critique of a Debate." *Critique of Anthropology* 5:69–96.

Kanō, H. (1980). "The Economic History of Javanese Rural Society: A Reinterpretation." *The Developing Economies* 18:3–22.

Keuning, S. (1984). "Farm Size, Land Use and Profitability of Food Crops in Indonesia." *Bulletin of Indonesian Economic Studies* 20:58–82.

Knight, G. (1982). "Capitalism and Commodity Production in Java." In Hamza Alavi, P. Burns, G. Knight, P. Mayer, and D. McEachern, eds., *Capitalism and Colonial Production.* London: Croom Helm.

Lyon, M. (1970). *Bases of Conflict in Rural Java.* Research Monograph no. 3. Berkeley: University of California Center for South and Southeast Asian Studies.

Mears, L. (1981). *The New Rice Economy of Indonesia.* Yogyakarta: Gadjah Mada University Press.

——— (1984). "Rice and Food Self-Sufficiency in Indonesia." *Bulletin of Indonesian Economic Studies* 20:122–138.

Meyer Ranneft, J. (1974 [1919]). "Het desawezen en het grondbezit in de afdeeling Cheribon" [Village and land tenure in Cirebon district]. In J. Meyer Ranneft, *Desa-Rapporten.* Jakarta: Arsip Nasional.

Meyer Ranneft, J., and W. Huender (1926). *Onderzoek naar den Belastingdruk op de Inlandsche Bevolking van Java en Madoera* [Inquiry into the tax burden on the native population of Java and Madura]. Weltevreden: Kolff.

Mortimer, R. (1972). *The Indonesian Communist Party and Land Reform, 1959–1965.* Monash University, Center of Southeast Asia Studies, Papers on Southeast Asia no. 1.

——— (1974). *Indonesian Communism Under Sukarno: Ideology and Politics 1959–1965.* Ithaca: Cornell University Press.

O'Malley, W. (1977). "Indonesia in the Great Depression: A Study of East Sumatra and Yogyakarta in the 1930s." Unpublished Ph.D. dissertation, Cornell University.

Palmer, I. (1976). "Rural Poverty in Indonesia with Special Reference to Java." Geneva: WEP Research Working Paper.

——— (1978). *The Indonesian Economy Since 1965: A Case Study in Political Economy.* London: Frank Cass.

Papanek, G. (1985). "Agricultural Income Distribution and Employment in the 1970s." *Bulletin of Indonesian Economic Studies* 21:24–50.

Pigeaud, T. (1960). *Java in the Fourteenth Century: A Study in Cultural History:*

The Nagara-Kertagama by Rakawi Prapanca of Majapahit, 1365 A.D. 5 vols. The Hague: Martinus Nijhoff.

Ploegsma, N. (1936). "Oorspronkelijkheid en Economisch Aspect van het Dorp op Java en Madoera" [Originality and economic aspect of the village in Java and Madura]. Dissertation, Leiden University.

Pudjiwati Sajogyo (1985). "The Impact of New Farming Technology on Women's Employment." In *Women in Rice Farming*. Aldershot: Gower/International Rice Research Institute.

Race, J. (1981). "The Political Economy of New Order Indonesia in a Comparative Regional Perspective." In J. Fox, R. Garnaut, P. McCawley, and J. Mackie, eds., *Indonesia: Australian Perspectives*. Canberra: Australian National University.

Raffles, T. (1978 [1817]). *The History of Java*. Oxford in Asia Historical Reprints, vol. I. Kuala Lumpur.

Ricklefs, M. (1981). *A History of Modern Indonesia*. London: Macmillan.

Rieffel, A. (1969). "The Bimas Program for Self-Sufficiency in Rice Production." *Indonesia* 8:103–133.

Robison, R. (1986a). "Into a New Phase Where Economic Laws Prevail." *Far Eastern Economic Review* 19 May 1986, pp. 28–30.

——— (1986b). *Indonesia: The Rise of Capital*. Southeast Asia Publications no. 13. Sydney: Allen & Unwin for the Asian Studies Association of Australia.

Roekasah, E., and D. Penny (1967). "Bimas: A New Approach to Agricultural Extension." *Bulletin of Indonesian Economic Studies* 7:60–69.

Schulte Nordholt, N. (1986). "From LSD to LKMD: Participation at the Village Level." In P. Quarles van Ufford, ed., *Local Leadership and Programme Implementation in Indonesia*. Amsterdam: Free University Press.

Soetrisno, L. (1981). "Selomartani Village Profile." In Mubyarto, ed., *Case Studies from Indonesia*. Nagoya: United Nations Center for Research and Development.

Soewardi, H. (1976). *Respons Masyarakat Desa Terhadap Modernisasi Produksi Pertanian Terutama Padi* [The response of village society to modernization of agricultural production, particularly rice]. Yogyakarta: Gadjah Mada University Press.

Sukarno (1970). "Marhaen, a Symbol of the Power of the Indonesian People." In H. Feith and L. Castles, eds., *Indonesian Political Thinking 1945–1965*. Ithaca: Cornell University Press.

Törnquist, O. (1984). *Dilemmas of Third World Communism: The Destruction of the PKI in Indonesia*. London: Zed Books.

Van der Kolff, G. (1937). *The Historical Development of Labour Relationships in a Remote Corner of Java as They Apply to the Cultivation of Rice*. New York: Institute of Pacific Relations.

Van Niel, R. (1964). "The Function of Landrent Under the Cultivation System in Java." *Journal of Asian Studies* 23:357–375.

——— (1980). "Measurement of Change Under the Cultivation System in Java, 1837–1851." In C. Fasseur, ed., *Geld en Geweten*. Vol. 1. The Hague: Nijhoff.

Wertheim, W. (1964). *Indonesian Society in Transition*. The Hague: van Hoeve.

White, B. (1983). 'Agricultural Involution' and Its Critics: Twenty Years After." *Bulletin of Concerned Asian Scholars* 15:18–31.

——— (1985). "Women and the Modernisation of Rice Agriculture: Some General Issues and a Javanese Case Study." In *Women in Rice Farming*. Aldershot: Gower/International Rice Research Institute.

Chapter Thirteen

Agrarian and Nonagrarian Bases of Inequality in Nine Javanese Villages

BENJAMIN WHITE AND GUNAWAN WIRADI

Between 1968 and 1973 Indonesia's Agro Economic Survey carried out regular surveys of fixed samples of farm households in twenty selected villages in the main rice-producing regions of Java. Summarizing the results of the first three seasons of these surveys (1968/69–1969/70), Sajogyo and Collier commented: "Although sample farmers have responded as well as local situations and incentives permitted in adopting the new high-yielding varieties, there is no picture yet of any 'green revolution' by Java's paddy farm operators" (Sajogyo and Collier 1972:43).

In 1981 the Agro Economic Survey research team had the opportunity to revisit nine of these twenty sample villages. By this time virtually all paddy cultivators were using modern rice varieties (MV hereafter), greatly increased inputs of chemical fertilizers and insecticides, and had achieved impressive yield increases, suggesting that in these villages as in the rest of Java Green Revolution production is now firmly and successfully established after an initially shaky start. Comparison of these two studies therefore offers an unusual opportunity to examine changes in agrarian economy and agrarian structure in the nine villages, during a period which has seen profound changes not only in agricultural production but also in state policies and in the organization of economic and political life at the local and national levels (cf. Hüsken and White, Chapter 12).

Multiple-village resurveys of this kind have both advantages and disadvantages compared to other types of research. The inclusion of nine villages, scattered across the three provinces of Java (see Map 4), permits us both to examine local variations (which, as we will see, are considerable) and also to draw some cautious conclusions about the general directions of change. However, the use of material of this kind for analysis of the processes and mechanisms of agrarian change presents many problems. First, neither the baseline nor the resurvey studies were designed specifically for this purpose. The baseline surveys consisted mainly of traditional farm management questionnaires; the 1981 resurveys, carried out as part of a series of Research Training Workshops on Land Tenure and Agrarian Relations, come closer to what is needed but still leave many gaps. Second, the inclusion of nine villages, while allowing more detailed analysis at community level than large-scale sample surveys, does not permit the depth of intensive single-village studies such as those of Hüsken, Anan Ganjanapan, and others in this volume, particularly with regard to qualitative material. While participant observation, informal interviews, and the writing of field notes have been a regular feature of the Agro Economic Survey's field research alongside questionnaire surveys in both periods, much of this qualitative information has not been preserved. Third, the sampling procedures differed in the baseline and resurvey studies, so that comparisons over time must be treated with caution.[1] Furthermore, we are dealing with two 'time-slice' studies separated by ten years rather than a systematic long-term study involving regular monitoring of changes; while it is possible to identify many changes during this period, it is rarely possible to document precisely when, why, or how they have occurred.

Given the data at our disposal, our analysis is restricted to the economic rather than the political, social, and ideological aspects of agrarian change. We first examine changes in paddy production and in the ways in which paddy incomes are distributed among the different landholding and landless classes; although many of the villages show a high degree of economic diversification, paddy is still the largest single source of incomes in all villages, with paddy incomes representing about two-thirds of all agricultural incomes and more than one-fifth of all incomes. Our analysis shows that although the agricultural sector by now generates incomes more or less sufficient to maintain the rural population at a minimal level of living, the great majority of these incomes are captured by a minority of larger landowning households ('large' in the context of Java meaning those owning more than 1.0 ha)—leaving the majority of households with farm and/or agricultural wage incomes which do not provide even minimal levels of reproduction. This picture of stark inequality and widespread poverty in conditions of agricultural productivity growth is altered, but only partly so, by the inclusion of nonfarm income sources in the analysis. Patterns of nonfarm income distribution tend to reflect, though in more muted form, the inequalities observed in the agricultural sector. Particularly for landless households they are a far more important source of income than agricultural sources. These patterns (which are subject to great variation between villages) provide grounds for questioning attempts to define rural 'classes' or the nature of rural 'classness' itself on the basis of agricultural production relations alone.

In attempting in a short paper to analyze agrarian conditions in nine villages at two points in time, we—and in turn our readers—face a difficult problem of presentation. Given the wide variations between villages, no useful purpose is served by lumping them together in 'average' form except for certain restricted purposes. We have therefore presented the quantitative data for each village separately, but in as abbreviated form as possible, in a series of tables covering all nine villages; this unavoidably requires straining the reader's patience and eyesight with a number of tables in 'cinemascope' form. For those allergic to such tables we have tried to make the text as intelligible as possible on its own account.

Finally, to avoid possible confusion, readers should note that while we speak of nine 'villages' in the analysis which follows, our household-level data are in fact derived from surveys of *parts* of villages, comprising one or more hamlets or 'neighborhoods' numbering less than 200 households (1971) and less than 150 households (1981), as described in note 1. 'Villages' in Java (i.e., the administrative units called *desa* or *kelurahan*) number normally several hundred and often over a thousand households; the following analysis describes patterns of change and inequality at 'neighborhood' rather than at *desa* level.

General Description of the Nine Villages

The sample villages are evenly distributed between the three provinces of Java and well dispersed within each province, each in different *ka-*

bupaten (districts) and at some distance from each other (see Map 4). Five of them lie in lowland plains close to sea level (among them, Wanarata is close to the coast and about one-fifth of its household heads are engaged in marine fishing); the remainder are in more elevated and hilly regions (see Table 13.1) but still within the range at which MV rice performs well. Columns 2–5 of Table 13.1 provide some information on the accessibility and the crude and agrarian population densities of the villages. Although all of the villages can be reached by motor vehicle, three of them have no regular public transport service, not lying on the routes of the small 'Colt' passenger vans or pickup trucks which have spread rapidly in rural Java during the 1970s. There is great variation in population and agrarian densities, as shown in columns 4–6 of Table 13.1: Mariuk in West Java, for example, has barely two-thirds and Janti (East Java) almost three times as many people per cultivated hectare as the all-Java average of 11. Variations in the availability of *sawah* (irrigated, or in the case of Sentul rain-fed, rice terraces) per household bear little relation to average operated holdings per cultivator household (columns 6–7), which may be three times larger than the all-household average in villages such as Mariuk and Rowosari where high rates of landlessness are found.

All of the above indicators of population pressure on land resources in turn do not seem closely related to average per-capita incomes, to the incidence of poverty (indicated in this case by per-capita incomes of households below 320 kg milled-rice equivalent per year), or to the proportion of household incomes derived from nonfarm sources (columns 8–10). The high proportions of average incomes derived from nonfarm sources (more than 50% in six cases, close to 50% in two cases, and 33% in the single case of Mariuk) serve as an important reminder that analyses of 'agrarian' differentiation, even in main rice-producing regions such as those represented by the sample villages, cannot limit themselves to the agricultural sector alone.

The most common cropping pattern on *sawah* in 1981 was simple double-cropping of paddy in all villages except one (Jatisari) which has achieved the more intensive pattern of five paddy crops each two years and another (Sentul) whose rain-fed *sawah* permits only a single paddy crop followed by fallow. Three villages (Kebanggan, Geneng, and Janti) lie within the catchment area of sugar factories and between 20 and 35% of their *sawah* is reserved in rotation for sugarcane cultivation as it has been since the colonial period. This land was formerly rented to the factories by its smallholder owners and its cultivation undertaken

TABLE 13.1. *General Data on the Nine Case-Study Villages: 1981*

Village	District	Elevation (meters above sea level) (1)	Distance to sub-district town (km) (2)	Access to public motor transport (3)	Population per square km (4)	Agrarian density (persons per cultivated ha) (5)	*Sawah* per household (ha) (6)	*Sawah* per cultivator household (ha) (7)	Average per capita income (Rp.'000/yr) (8)	Households below poverty line[a] (9)	Total income from nonfarm sources (10)
West Java											
Sentul	Serang	25	2	+	812	9	0.44	0.57	57	75%	67%
Mariuk	Subang	10	8	–	658	7	0.53	1.52	124	44%	33%
Jatisari	Cianjur	350	6	+	865	10	0.37	0.51	140	40%	48%
Central Java											
Rowosari	Kendal	8	6	–	1437	19	0.19	0.60	101	49%	69%
Kebanggan	Banyumas	325	7	+	1481	15	0.24	0.59	102	46%	59%
Wanarata	Pemalang	312	7	–	413	10	0.21	0.33	87	54%	54%
East Java											
Sukosari	Jember	312	4	+	1300	17	0.37	0.74	205	25%	55%
Geneng	Ngawi	53	2	+	1230	15	0.25	0.54	148	22%	45%
Janti	Sidoarjo	7	4	+	1842	31	0.12	0.31	199	13%	63%

Source: Resurveys, 1981.
[a] Households with per capita income in 1981 below 320 kg milled-rice equivalent per year at local prices (approximately Rp. 62,500 per capita).

by the factories using wage labor, but since the introduction of the new TRI or TRIS (smallholder sugarcane intensification) programs in the late 1970s sugarcane is now cultivated by groups of smallholders themselves and sold to the factories.

Changes in Paddy Technology and Productivity: 1971–1981

Given the long history of double-cropping in most of the villages, there have been relatively few increases in cropping intensity. The major thrust of intensification in all villages has been in raising paddy yields through increased use of modern inputs. The use of either 'national improved' or 'modern' varieties (the latter being those originally diffused from IRRI) was already quite widespread in many villages by 1971, as may be seen in Table 13.2. In all the irrigated villages (i.e., all except Sentul) even greater numbers of farmers were already using chemical fertilizers (column 4); a few farmers in some villages had used them in small quantities since the late 1950s and in the occasional case since the 1930s (Agro Economic Survey 1972). Thus, between 1971 and 1981 changes in modern input use, though considerable, have been quantitative rather than qualitative. By 1981 all paddy farmers were using MVs and furthermore planting them on all the *sawah*, excepting only the occasional small plot reserved for good-tasting or glutinous (*ketan*) varieties for use on special occasions; the single exception is Jatisari, in which about 10% of farmers continue to produce local varieties and a single crop, a portion of upland terraces in this region being officially designated for the traditional 'Cianjur' varieties which fetch high prices in urban markets. Fertilizer use is now also universal (column 5) and dosages have doubled or more than doubled in many villages (columns 6–7), reaching the high levels of more than 400 kg/ha in Janti and Geneng.

These changes, together with insecticide use, the shift in the late 1970s to pest-resistant MVs, and associated Green Revolution practices such as straight-row planting and more frequent weeding, have resulted in the marked yield increases which can be seen in Table 13.2 (columns 8–9. These increases on the one hand are remarkable evidence of the ingenuity of Java's farmers—despite small farm sizes, widespread illiteracy and the uneven record of the various agencies whose task is to serve them—in the application of science in the process of production. This applies to both smaller and larger farmers, to both owner-operators

TABLE 13.2. *Changes in Paddy Cultivation Practices and Yields: 1971–1981*

Village	Paddy farmers using new varieties			Paddy farmers using chemical fertilizer		Average fertilizer dosage (kg/ha)		Average paddy yields (t/ha)	
	1971, NIV (1)	*1971, MV* (2)	*1981, MV* (3)	*1971* (4)	*1981* (5)	*1971* (6)	*1981* (7)	*1971* (8)	*1981* (9)
West Java									
Sentul	87%	3%	100%	19%	100%	n/a	n/a	1.1	2.0
Mariuk	23%	6%	100%	89%	100%	135	257	2.5	3.8
Jatisari	3%	23%	93%	81%	100%	277	350	3.4	4.3
Central Java									
Rowosari	24%	6%	100%	54%	100%	58	205	3.5	4.6
Kebanggan	—	22%	100%	96%	100%	174	327	2.8	3.5
Wanarata	92%	62%	100%	78%	100%	121	392	2.9	3.0[a] 4.3[b]
East Java									
Sukosari	2%	96%	100%	84%	100%	n/a	n/a	3.0	4.7
Geneng	1%	54%	100%	96%	100%	176	574	3.1	5.2
Janti	1%	11%	100%	92%	100%	213	444	2.7	5.2

Sources: Agricultural census, 1971 (cols. 1, 2, 4); Sample survey, 1971 (cols. 6, 8); Resurvey, 1981 (cols. 3, 5, 7, 9).
Note: '1971' and '1981' refer to wet seasons 1970–1971 and 1980–1981. NIV = national improved varieties; MV = modern varieties; n/a = not available.
[a] Actual yield including harvest failure.
[b] Estimated normal yield.

and tenants (technology adoption and yields are not markedly influenced by the form of land tenure, cf. Sajogyo and Collier 1972:14), and to both men and women (the latter being important decision-makers in agricultural production, cf. White and Hastuti 1980). On the other hand, the single most important economic factor promoting yield increases has undoubtedly been state pricing policies, in particular the massive fertilizer subsidy which since 1971 has roughly halved the real price of fertilizer. In 1971 farmers in the nine villages paid between 1.16 and 1.65 kg paddy equivalent for each kilogram of fertilizer, while in 1981 they paid only between 0.61 and 0.75 kg (Masjidin Siregar and Aladin Nasution 1984:135).

Mechanization in paddy cultivation has occurred in only two villages. In Mariuk, the area of *sawah* prepared with tractors had risen from a very small proportion in 1971 to more than 90% in 1981. This development seems related to the stricter cultivation schedules (shortening the land-preparation period) resulting from the improved irrigation system already mentioned. In Geneng, tractors are a more recent innovation and coincided with the introduction since 1978 of water pumps for dry-season cultivation; these pumps like the tractors are purchased by larger farmers (in most cases, with subsidized bank credit or dealer credit) and used both on their own fields and for hire to other farmers. Tractors in this village were at first used solely for land preparation and for the first plowing, while animal power was used for second plowing, harrowing, and puddling. After some years, as government promotion of uniform planting schedules (for purposes of water rotation and crop protection) made the cultivation season shorter, tractor owners found their profits from tractor hire also shrinking and began to use the engines to power irrigation pumps, which analysis shows to be more profitable (Masjidin Siregar and Aladin Nasution 1984).

Other technological changes universal in the nine villages, which like tractors affect not so much productivity as labor absorption, include the shift from hand weeding to the use of rotary or toothed weeding implements (made possible by straight-row planting), the almost universal shift from finger-knife (*ani-ani*) harvesting to sickles, and in paddy processing the rapid spread of small diesel-powered rice-hullers which began in the late 1960s and by the mid-1970s had almost completely replaced hand pounding, previously an important source of income for women of poorer households.

Changes in the Paddy Earnings of Farmers and Hired Labor

We will first examine changes in the quantities and shares of paddy incomes accruing to 'farmers' on the one hand and 'hired labor' on the other during this period of substantial productivity growth, before discussing changes in land tenure, labor use, and labor arrangements which underlie these changes. Table 13.3 shows how much of the total available income from paddy production in 1971 and 1981 (after deducting all nonlabor production costs) went into the pockets of farmers and hired laborers respectively, in seven of the case-study villages. (For Sentul and Sukosari these data are not available.) 'Farmer's share' in this case means a Chayanovian bundle of 'returns to land, family labor, management, and capital' (i.e., without the fictive separation of these into imputed land rents, interest, and family-labor wages), and 'hired labor's share' is the total actual wages paid out in cash or kind per hectare. In 1981, the farmer's share from each hectare of paddy production was generally more than 60% of production and hired labor's share generally less than 30% (the remaining roughly 10% representing nonlabor input costs, which have proportionately declined since 1971 with the help of fertilizer subsidies); in 1971 the ratio was more favorable to hired labor in all villages except Geneng (columns 1–4). Analyzing the same data from another angle (columns 5–7), it is clear that in all villages except Jatisari, the absolute quantity of (real) wages paid out per hectare has increased; in all villages except Geneng, however, the farmer's paddy income per hectare has increased more, and hired labor's less, than paddy production. These data therefore indicate a growing divide between 'farmers' on the one hand, and 'hired laborers' on the other, in terms of their relative ability to command incomes from paddy production. In practice, as we shall see below, many smaller farm households also hire out their labor and thus acquire income from both these sources.

In later sections we will examine changes in hired labor use, wage rates, and labor arrangements which influence the incomes of hired workers in paddy production. First, however, we will discuss the evidence of changes in patterns of land control, since these determine both the distribution of paddy incomes among farm households and also the separation between those who have access to a 'farmer's or landowner's share' and those landless households whose only potential access to incomes in this sector is in the form of a wage.

TABLE 13.3. *Changes in Paddy Income Shares (per Hectare) of Farmers and Hired Laborers: 1971–1981*

Village	Shares of total production received, 1971 (%)			Shares of total production received, 1981 (%)			Percentage change 1971–1981		
	Farmer (1)	*Hired labor* (2)	*Other inputs* (3)	*Farmer* (4)	*Hired labor* (5)	*Other inputs* (6)	*Yield per ha* (7)	*Farmer share* (8)	*Hired labor share* (9)
West Java									
Sentul	—	—	—	—	—	—	+92	—	—
Mariuk	53	36	11	56	31	13	+51	+62	+27
Jatisari	52	37	11	66	26	8	+25	+61	−13
Central Java									
Rowosari	60	32	8	61	29	10	+31	+34	+20
Kebanggan	60	27	13	62	24	14	+27	+34	+15
Wanarata	56	27	17	65	22	13	+49	+73	+24
East Java									
Sukosari	—	—	—	—	—	—	+57	—	—
Geneng	49	34	17	45	37	18	+66	+55	+77
Janti	43	35	22	61	27	12	+91	+168	+49

Source: Sample survey 1971; Resurveys, 1981.

Land Tenure and Land Markets: 1971–1981

Our only information on access to land in 1971 concerns the distribution of operated holdings (Agro Economic Survey 1972) without any details on landownership or tenure status. The comparison of operated holdings distribution in 1971 and 1981 shown in Table 13.4, however, suggests some tentative conclusions regarding ownership and tenure which can be further examined using the more detailed information available for 1981. The average area of *sawah* per operator household has apparently increased significantly in all villages except Wanarata (where at the time of the 1981 survey much *sawah* was not cultivated due to severe rat infestation problems) and a slight decline in Janti. How farm sizes can become larger when availability per household in a growing population becomes smaller is more easily understood when we note the increases in all villages (in many cases, very large increases) in the percentage of households not cultivating any *sawah* at all (columns 2 and 7)—i.e., 'landless' in one narrow sense of the term. These increases, together with the uneven farm-size distribution among operator households (columns 8–12) produce the rather high levels of overall inequality in farm-size distribution found in 1981: in six of the nine villages, one-half or more of households do not operate any *sawah*, and in five villages smaller groups of households with farms of more than 1.0 ha between them cultivate one-half or more of all available *sawah*.

Operated farm-size distribution is not a great deal more equal than ownership distribution, which is examined in Table 13.5. Again we find generally large proportions of 'landless' (in the narrow but different sense of nonowners)—one-half or more of households in six of the nine villages—and correspondingly large proportions of land owned by small groups, generally less than 10% of households, with holdings of 1.0 ha and above, which between them own more than half (and in the case of Mariuk, almost 90%) of all *sawah*. In five villages, at least one owner has *sawah* in excess of the (5 ha) maximum permitted by the 1960 Agrarian Laws; the single largest owners in these nine sampled neighborhoods of about 120 households (see note 1) generally own at least 10% of all *sawah*.

The overall similarity of ownership and farm-size distributions in fact goes together with considerable rates of land transfer through tenancy in the sample villages, two phenomena which at first sight appear con-

TABLE 13.4. *Changes in Size Distribution of Operated* Sawah *Holdings: 1971–1981*

Village	1971 Percentage of households by size group (ha)					1981 Percentage of households by size group (ha)					
	Average per operator household (ha) (1)	*0* (2)	*0.5* (3)	*0.5–0.99* (4)	*≥1.0* (5)	*Average per operator household (ha)* (6)	*0* (7)	*0.5* (8)	*0.5–0.99* (9)	*≥1.0* (10)	*Percentage of all* sawah *in holdings over 1.0 ha* (11)
West Java											
Sentul	0.51	5	46	34	15	0.57	24	35	28	13	41
Mariuk	1.18	33	19	15	33	1.49	64	9	7	20	87
Jatisari	0.46	22	52	14	12	0.51	26	50	12	12	51
Central Java											
Rowosari	0.22	48	43	6	3	0.60	69	20	6	5	51
Kebanggan	0.40	26	48	13	14	0.59	60	29	8	4	50
Wanarata	1.06	1	36	26	36	0.38[a]	37	47	11	5	26
East Java											
Sukosari	0.61	42	40	8	10	0.74	50	38	3	9	76
Geneng	0.40	38	34	16	11	0.54	53	27	17	3	22
Janti	0.35	52	40	6	2	0.31	61	35	3	1	22

Sources: Agricultural census, 1971; Resurveys, 1981.
[a] Much land was uncultivated in 1981 due to rat infestation.

TABLE 13.5. Sawah *Ownership Distribution: 1981 (Wet Season 1980–1981)*

Village	Percentage of households by size group (ha)					Percentage of all *sawah* owned by owners of more than 1.0 ha	Holding of single largest owner (ha)	Percentage of total
	0 (1)	*0.25* (2)	*0.25–0.49* (3)	*0.50–0.99* (4)	*≥1.0* (5)	(6)	(7)	(8)
West Java								
Sentul	30	16	17	27	10	37	2.8	7
Mariuk	70	1	6	5	18	89	11.9	21
Jatisari	32	28	15	12	13	55	2.6	5
Central Java								
Rowosari	64	10	15	5	6	66	3.3	10
Kebanggan	58	18	11	9	4	58	6.5	16
Wanarata	28	30	26	10	6	37	5.0	11
East Java								
Sukosari	50	29	9	4	8	74	10.7	12
Geneng	60	2	12	18	8	52	5.3	11
Janti	56	5	31	3	5	36	3.5	12

Source: Resurveys, 1981.

TABLE 13.6. *Tenure Status of Sawah* (1981) and Changes in Real Land Rents (Wet Season 1971–1981)

Village	Percentage of all *sawah* operated by			Real rent (tons paddy/ha/season)			Mortgage (*gadai*) frequency 1981	
	Owner (1)	*Lessee (fixed rent)* (2)	*Share tenant* (3)	*1971* (4)	*1981* (5)	*% change* (6)	*% of* sawah (7)	*% of households* (8)
West Java								
Sentul	62	—	38	—	—	—	18	24
Mariuk	83	4	13	1.4[a]	1.7[a]	+22	2	6
Jatisari	77	4	19	1.2	1.8	+52	2	13
Central Java								
Rowosari	72	26	2	1.3	1.6	+30	—	—
Kebanggan	52	43	5	1.0	1.3	+22	—	—
Wanarata	83	1	16	1.7	2.1	+25	2	7
East Java								
Sukosari	64	35	1	0.7	1.2	+82	10	11
Geneng	50	36	14	2.0[a]	2.7[a]	+35	—	—
Janti	71	29	—	1.9[a]	2.8[a]	+47	—	—

Sources: Sample survey, 1971; Resurveys, 1981.
[a] Rent per year.

tradictory. As may be seen in Table 13.6 (columns 2 and 3), while an overall majority of *sawah* is operated by its owners, substantial amounts varying from 17% in Wanarata and Mariuk to around 50% in Kebanggan and Geneng are operated under some tenancy arrangement; fixed-rent leasing is predominant in five villages, share tenancy in four, and in only one village (Geneng) are both forms significant. Further examination of the tenure status of households (Table 13.7) shows that despite the high proportions of land held in tenancy and of households involved in some tenancy transaction, many of these tenants are also owners (columns 3–5) so that tenancy transfers do not greatly reduce the proportions of 'absolute' landless, with the exception of Sentul and Geneng (compare columns 6–8 and column 9).

Fixed-rent leasing is quite common in the case of village officials' salary lands (*sawah bengkok*), which are found only in the six Central and East Java villages; this practice often reflects the officials' need for ready cash to repay various debts and promises incurred in their election campaigns. Real rental rates have substantially increased between 1971 and 1981 in all villages (Table 13.6, columns 4–6), although these increases have generally been lower than the yield and 'farmer's share' increases already seen in Table 13.3. Five different forms of fixed-rent leasing were found in the sample villages. In three of them (*sewa tahunan*, *jual oyodan*, and *jual potongan*) rent is paid in advance and thus limited to prospective tenants who have cash in hand; the two less common forms involving payment after harvest (*kontrak* and *setoran*) are found only in Jatisari and Kebanggan, and then only between kin (Gunawan Wiradi and Makali 1984).

For those with ready cash, another opportunity to expand control of land is through mortgage (*gadai*), which although banned in a regulation of 1960 is found in five villages (Table 13.6, columns 7–8). In this case the owner in need of cash receives more than in a fixed-rent transaction and often continues to work as a share tenant on the same land, continuing to bear the risk of cultivation (which perhaps explains its relative frequency in Sentul and Wanarata); on the other hand, the difficulties encountered in repayment often mean that *gadai* is the first step towards eventual loss of ownership rights. In some cases, small owners in need of cash (for investment, ceremonial expenses, children's school fees, etc.) are unable to lease out land because of its low productivity (as in Sentul, which has the highest *gadai* rates in our sample) and mortgage is the only choice. In other cases a rental market exists (Kebanggan, Sukosari),

TABLE 13.7. Sawah *Tenure Status of Households (Percentage of All Households): 1981 (Wet Season)*

Village	*Sawah* owners					Nonowners				Percentage of all households involved in tenancy transfers[a]
	Non-operators (1)	*Own land only (2)*	*Own and lease (3)*	*Own and sharecrop (4)*	*Own and lease and sharecrop (5)*	*Lease only (6)*	*Sharecrop only (7)*	*Lease and sharecrop (8)*	*'Absolute' landless (9)*	(10)
West Java										
Sentul	9	48	—	13	—	—	16	—	14	38
Mariuk	4	25	—	—	1	2	7	—	62	14
Jatisari	3	57	1	8	—	2	5	1	23	20
Central Java										
Rowosari	13	18	5	—	—	7	—	1	56	26
Kebanggan	10	25	6	1	—	7	1	—	50	25
Wanarata	11	56	—	5	—	1	1	—	26	7[b]
East Java										
Sukosari	1	43	5	1	—	1	—	—	49	8
Geneng	14	19	5	2	—	14	5	2	39	42
Janti	9	33	2	—	—	4	—	—	52	15

Source: Resurveys, 1981.
[a] Col. 10 = cols. 1 + 3 + 4 + 5 +6 +7 +8.
[b] Excludes the 11% noncultivating owners in col. 1, most of whom left their *sawah* uncultivated in 1981.

but those needing cash may prefer *gadai* because the initial payment is higher and also because they still have access to a reduced income from their land as sharecroppers or (in Sukosari) by an informal guarantee of wage employment. Not all *gadai* transactions, however, involve hard-pressed smallholders and wealthy mortgagors. In Mariuk and Jatisari, land is commonly released in *gadai* by relatively large owners as part of an expansion strategy, sacrificing part of their land for a number of years in order to purchase more land with the proceeds and using the income from the newly purchased land to reacquire the mortgaged portion (Gunawan Wiradi and Makali, 1984).

In all the eight villages where share tenancy is found, the tenant receives 50% of the gross yield. In each village, however, at least two and sometimes (Jatisari and Wanarata) as many as four different arrangements are found for the division of production costs (seeds, fertilizers, animal/tractor power, labor, land tax, irrigation fees) between owner and tenant. Universally, all labor costs and (where applicable) tractor or animal hire are paid in full by the share tenant, despite the 1959 Share Tenancy Regulations which stipulate the sharing of all production costs in the same proportion as the sharing of the product. Modern inputs such as chemicals are also paid in full by the tenant with the exception of a minority of cases in Jatisari; seeds are also provided by the tenant in all cases in most villages, and in the majority of cases in Sentul, Jatisari, and Wanarata. The landowner, however, more commonly pays all or part of land taxes and irrigation fees (Gunawan Wiradi and Makali 1984:table 4.12, appendix 4.6). In Mariuk and Jatisari, those who wish to obtain share tenancies commonly work first for some time as a permanent laborer or 'apprentice' (*ngawula* or *magang*) for the landowner. In Mariuk, also, wage labor in the wet season is commonly a precondition for obtaining a share tenancy on part of the employer's land in the dry season (cf. the similar arrangement described in Thailand by Anan in Chapter 5). Large landowners in this way ensure themselves of a wet-season labor supply—even in these crowded conditions and with widespread landlessness, labor recruitment can be a problem at peak periods, as we will see in the next section—and also commonly tie these laborers more securely by keeping them supplied with loans.

This brief summary of the complex and heterogeneous patterns of tenancy in the sample villages—heterogeneous not only in form but also in function—suggests that fixed-rent transactions occur mainly between landed households and with rented land flowing from smaller to larger owners or between large owners; share-tenanted land, on the other hand, will generally flow from larger owners to smaller owners and the landless. This pattern of countervailing flows of fixed-rent and share tenancies, and the frequency of 'horizontal' flows of fixed-rent tenancies within the landowner classes, is confirmed by detailed analysis of tenancy flows in Rowosari and Kebanggan (Suseno et al. 1982; Retno Setyowati et al. 1982); under such conditions, even high tenancy rates do not make the distribution of operated holdings differ greatly from the distribution of landownership.

Little information is available from the 1981 surveys on the mechanisms which have produced the large increases in the proportions of households without paddy farms which we have seen in Table 13.4. Since the proportions of noncultivating owners (cf. Table 13.7) are not large, these increases must reflect either the loss of land through sale, the loss of tenancy rights (particularly share tenancies if the preceding arguments are correct), or increasing numbers of 'new' households who because of the absence or small size of parental holdings did not (or did not yet) inherit land. It is likely that all three mechanisms have been at work.

The various individual case-study reports from the sample villages give evidence of a lively and growing land sale market in all villages, although it was not always possible to quantify land sales. In Rowosari, Kebanggan, Wanarata, and Sukosari data are available on the ways in which *sawah* presently owned was acquired by households in the different landowning classes. While the great majority of land owned by smaller owners came to them through inheritance, owners of more than 0.5 ha tend to have acquired large proportions of their land through purchase; the great majority of land sales have transferred land to these groups from smaller owners (Djoko Kustiono et al. 1982; Retno Setyowati et al. 1982; Suseno et al. 1982; Umar Wahyu Widodo et al. 1982). On the other hand, generally less than 10% of presently nonowning households in the nine villages report having formerly owned land and having lost it through sale. It appears, then, that the resumption of share-tenanted land from former tenants, by owners wishing to cultivate it themselves or to shift to fixed-rent leasing, has been a more important factor than losses through sale in the rapid growth of noncultivator 'absolute' landless households. Our information on this point is incomplete, but there is clear evidence in some villages of a decline in share tenancies, and landless respondents in many others report increasing difficulty in obtaining share tenancies.

Changes in Hired Labor Use, Wage Rates, and Labor Arrangements

In the previous section we have observed the growth in the proportion of households who, being neither owners nor cultivators of *sawah*, have potential access to paddy incomes only in the form of wages. Earlier, we noted that as paddy production has increased, the proportion of paddy income per hectare paid out in the form of wages (i.e., the income available for division among this growing group) has generally declined although in most cases it has absolutely increased. These trends might reflect changes in the amount of hired labor used per hectare, changes in wage rates for specific forms of labor, changes in the mode of hired labor recruitment and payment, or some combination of these.

Table 13.8 compares preharvest labor use per hectare (harvest labor being almost impossible to estimate reliably) in the wet seasons 1970–1971 and 1980–1981. In four villages total labor use per hectare has remained virtually stable, in two (Jatisari, Rowosari) it has sharply declined, and in two (Kebanggan, Geneng) it has sharply increased. Hired labor use on the other hand has declined in five villages and increased in only two. (We exclude Sentul, where the transition from exchange to hired labor has only recently begun, with only 14% of paddy farmers using any hired labor at all in 1981.) Since the demand for transplanting labor (a female task) is not affected by new technology, the reasons for the decline may be mainly sought in labor use and technology in land preparation, the main male task in preharvest work. As shown in Table 13.8, tractor use in Geneng has displaced both hoeing labor and draft cattle (and the latter's hired operators) while in Mariuk it has displaced plowing but made little inroads on hoeing. In Jatisari, Rowosari, and Kebanggan hoeing has been displaced not by tractors but by the increased use of draft cattle. A major part of the decline in hired labor use, however, is due to the increased inputs of family labor in most villages. (Here we should ignore Wanarata, whose temporarily reduced farm sizes in 1981 have caused an abrupt shift to greater use of family labor.) This can occur even where farm sizes increase because of the greater demands of intensified production for small tasks which (unlike land preparation and transplanting) do not require mass labor inputs at

TABLE 13.8. *Changes in Labor Use in Paddy Cultivation: 1971–1981 (Person-hours per ha per Crop, Preharvest, Wet Seasons 1970–1971 and 1980–1981)*

	Person-hours/ha/crop								Changes in land preparation labor use 1971–1981 (%)		
Village	*Total 1971 (1)*	*Total 1981 (2)*	*Family 1971 (3)*	*Family 1981 (4)*	*Hired 1971 (5)*	*Hired 1981 (6)*	*% Change, total (7)*	*% Change, hired (8)*	*Men (9)*	*Cattle (10)*	*Tractors (11)*
West Java											
Sentul[a]	458	439					−4				
Mariuk	692	686	65	113	627	573	−1	−9	−14	−97	+
Jatisari	1702	1263	134	161	1568	1102	−26	−30	−38	+44	−
Central Java											
Rowosari	1352	808	87	48	1265	760	−40	−40	−27	+8	−
Kebanggan	828	1075	56	191	772	884	+30	+14	−10	+51	−
Wanarata[b]											
East Java											
Sukosari	713	689	70	193	643	496	−3	−23	n/a	n/a	−
Geneng	821	781	54	128	767	652	−5	−15	−49	−65	+
Janti	757	1105	104	254	653	821	+46	+26	+3	−8	−

Sources: Sample surveys, 1971; Resurveys, 1981 (Faisal Kasryno 1984:table 6.6; Masjidin Siregar and Aladin Nasution 1984:table 5.5).

[a] Hired labor was not found in Sentul in 1971 and was used by only few farmers in 1981.

[b] Temporarily reduced farm sizes in Wanarata render comparisons inappropriate in this case.

one time—for example, fertilizer and pesticide application, crop inspection, and water control, all of which can be undertaken by family members.

Under these conditions of generally stagnant or declining demand for hired labor, together with growth in landlessness, what has happened to wage rates? This issue is rather complex due to the coexistence in most villages of several forms of hired labor, each with its own mode of payment. Table 13.9 shows first the changes in real wage rates for casual daily labor in the main preharvest tasks of plowing/harrowing and hoeing (both male) and transplanting and weeding (the first always, and the second predominantly, female) and also (columns 4–6) the wages paid out to harvesters. In the case of the preharvest wages shown in columns 1–3, we observe a general trend without exceptions, however hard it may be to explain: real wage rates have increased in all villages for both male and female labor and in some cases have even doubled. The method of measuring cash wages against a local rice price has resulted in some overestimation of these increases, since in recent years rice prices have risen more slowly than those of other basic commodities such as cloth (cf. Faisal Kasryno 1984:184); unfortunately no weighted index of rural consumer prices is available as far back as 1971. However, adjustment would not alter the conclusion that real wages have generally increased. Having no information from the years between 1971 and 1981 we cannot say precisely when this increase has occurred; data from an Agro Economic Survey wage-monitoring project in other sample villages (Mazumdar and Husein Sawit 1986) as well as available large-scale data (Papanek 1985) support the general impression provided by respondents in our nine sample villages that real wages were largely stagnant during the 1970s and did not begin to increase markedly until about 1980, i.e., in the last two years of the period in question.

Similar increases have occurred in harvesting wages (columns 4–6), which account for a large proportion (generally around one-third) of all paid-out wages in paddy cultivation; these are harder to determine precisely because harvesters are generally paid with an in-kind proportion (*bawon*) of the amount they harvest, and also because several different *bawon* rates are found simultaneously in all villages except Sukosari, depending on the social or kin relations between employer and harvester. Columns 4 and 5 of Table 13.9 show the downward proportional shift of *bawon* payments between wet seasons 1968–1969 (the data are not available for 1970–1971) and 1980–1981, confirming re-

TABLE 13.9. *Changes in Real Wage Rates for Casual Daily Labor (Wet Seasons 1970–1971 and 1980–1981) and Wages Paid to Harvesters*

Village	Changes in real wage rate[a] for casual daily labor 1971–1981			In-kind harvest wages paid out per hectare		
	Plowing/ harrowing (men) (1)	*Hoeing (men)* (2)	*Transplanting/ weeding (women)* (3)	*% of gross yield 1968–1969*[b] (4)	*% of gross yield 1980–1981* (5)	*Quantity change* (6)
West Java						
Sentul		(- - - Exchange Labor - - -)		17	15	+71%
Mariuk	+124%	+25%	+24%	15	12	+76%
Jatisari	+41%	+19%	+11%	10	9	+56%
Central Java						
Rowosari	+91%	+80%	+21%	10	9	+57%
Kebanggan	+18%	+28%	+65%	19	19	+70%
Wanarata	+46%	+12%	+41%	10	9	+43%
East Java						
Sukosari	+6%	+73%	(*kedokan*)	20	20	+214%
Geneng	+106%	+119%	+124%	16	11	+112%
Janti	+45%	+78%	+104%	18	17	+92%

Sources: Sample surveys, 1969 and 1971; Resurveys, 1981 (Gunawan Wiradi and Makali 1984:tables 4.16–4.18, 4.20, and app. 4.4).

[a] The cash component of the wage (excluding value of meals, etc.) converted to paddy equivalent at local prices.

[b] Data for 1970–1971 are not available.

ports by many authors. Column 6, however, shows that this proportional decline has been more than compensated by yield increases, so that the total quantity paid out to harvesters has markedly increased in all villages.

These increased wage rates are puzzling when viewed in the context of rapidly increasing landlessness and general declines in hired labor use. The most likely explanation, based on qualitative information from the sample villages, seems to lie in the increasingly strict cultivation schedules being practiced in recent years. As the demands of water rotation and pest control become heavier under conditions of intensive cultivation, farmers have been urged and in some cases obliged to plant at the same time, and they also now tend to plant a restricted range of modern varieties with more uniform maturation periods. They thus find themselves not only plowing, harrowing, hoeing, and planting but also weeding and harvesting at the same time as other farmers in the locality—i.e., facing more uniform, briefer, but higher peaks of labor demand which create short-term labor shortages at the local level. This interpretation is consistent with the shift to tractors in two villages and increased use of draft cattle in place of hoeing in four others (even though in three of these hoeing wages are becoming relatively cheaper than plowing wages, as we have seen in Table 13.8). The growth in seasonal peaks in labor demand may result in peculiar problems of adjustment in labor markets and in the division of labor, as was observed in Geneng where the highest growth in both preharvest and harvest wage rates has occurred; at the time of the 1981 resurvey fieldwork, when men were busy harvesting one paddy crop with sickles (having replaced female *ani-ani* harvesters some years before), women were for the first time seen hoeing in the *sawah* to prepare it in time for the following planting, a task which has previously been considered exclusively male in this village and throughout Java.

Whether these increased real wage rates, essentially anomalous in conditions of overall labor surplus, will continue or whether they represent a short-term reflection of changing labor-demand schedules which employers will find ways to overcome, are questions which our data cannot answer. However, another development consistent with the 'labor demand' hypothesis is the spread in many villages of alternative labor recruitment practices in place of casual daily labor which function for the employer either to assure him or her of labor supply at peak periods, to reduce labor costs as 'normal' casual wage rates rise, or in some cases to achieve both of these. Table 13.10 shows the frequency of various such practices in the sample villages in 1981. *Borongan* (contract) labor is now found in all except one village in preharvest work, and while it is not completely new there is general agreement that it is increasing. Individuals or more commonly groups of laborers are paid a fixed rate to complete some operation (land preparation, transplanting, or harvesting) on a certain area of *sawah*; the work is more quickly completed with laborers often working longer hours per day, sometimes far into the evening after working elsewhere for a 'daily' wage, and the total cost per hectare is less than casual daily labor (Soentoro et al. 1982:40). A smaller number of *borongan* workers therefore obtain a lower wage for the task but by expending more time or effort they can earn more each day.

In Kebanggan and Sukosari, virtually all farmers use the labor-tying arrangement known most frequently as *ceblokan* in West Java and *kedokan* in Central and East Java, although it also goes by a variety of other local names. In order to gain access to a harvest *bawon*, workers are obliged to perform some preharvest labor and sometimes also postharvest labor without pay. In Sukosari *kedokan* workers (called *betonan* workers if they make a group contract) must transplant, weed, harvest, and thresh the harvested paddy and cut the stubble in order to receive their one-fifth 'harvest' *bawon*, and in Kebanggan (where the system is called *paculan*) they perform all these tasks plus water control, hoeing, and the first drying of the threshed paddy for a *bawon* of one-fifth or one-sixth. The consequence of these arrangements (which, given the nature of the tasks, often link men and women in a single labor arrangement) is twofold. First, of course, there is more work for the same wage (although, as pointed out earlier, bigger harvests may increase the absolute value of the *bawon*); but secondly, various periods in the cropping cycle that were formerly times of immediate cash income in daily wage employment (land preparation, transplanting, weeding) now become periods of unpaid work and therefore a time when the probability of having to seek a consumption loan increases. The natural source of such a loan is the *kedokan* employer, who 'owes' the laborer a harvest wage and can therefore recoup the loan and interest by deduction at source when the *bawon* wage is finally paid, thus solving debt-collection problems.

In Rowosari and Wanarata the majority of paddy farmers now arrange their harvest by selling the standing crop to a middleman (*penebas*)

TABLE 13.10. *Modes of Labor Recruitment and Payment in Paddy Cultivation: 1981 (Wet Season 1980–1981), Use by Percentage of Employers*[a]

Village	Exchange labor (1)	Casual daily labor (2)	Contract labor (3)	*Kedokan* (4)	Harvest *bawon* (5)	*Tebasan* harvest (6)	Open/restricted harvest[b] (7)
West Java							
Sentul	86	14	5	—	100	—	R
Mariuk	—	88	62	—	100	—	O
Jatisari	—	80	28	—	100	—	O
Central Java							
Rowosari	—	91	24	—	12	88	O/R
Kebanggan	—	93	—	90	100	—	R
Wanarata	—	94	21	—	3	97	O/R
East Java							
Sukosari	—	84	25	100	100	—	R
Geneng	—	100	44	—	100	—	O
Janti	—	100	100	—	100	—	R

Source: Resurveys, 1981 (Gunawan Wiradi and Makali 1984:tables 4.15, 4.19, and app. 4.4).

[a] Since many employers use more than one form of recruitment, row percentages often total more than 100.

[b] O = open to all comers;
R = restricted access (invitees or those who have done prior unpaid work);
O/R = open 1971, restricted 1981.

shortly before harvest. The *penebas* brings his or her own team of harvesters, using sickles, who are paid a piece-rate cash wage in place of the traditional *bawon*. The *tebasan* system, which was already used by smaller numbers of farmers in 1969–1970, represents a reduction in the effective harvest share from the 'traditional' one-eighth (which in practice often approached one-sixth through skillful selection of the biggest bundles by harvesters for their own share) to one-eleventh (Collier et al. 1974). As Table 13.10 indicates, there remain only three villages where harvest labor opportunities are open to all comers, and in all villages some form of restrictive or 'exclusionary' labor recruitment arrangement (cf. Hart 1986) is now practiced; while none of these forms are new in Java (Hüsken and White, Chapter 12), there is little doubt that they are increasing.

Agricultural and Nonagricultural Bases of Inequality and Differentiation

The preceding sections have documented some of the variations between villages in the precise form and pace of agrarian changes occurring during a decade of successful agricultural intensification. Despite these variations some broad generalizations are possible. We have seen that increasing landlessness now leaves about half of all households without *sawah* ownership rights and about 40% without cultivation rights. The decline in share tenancies has contributed to the relative decline of smaller farm households (although their absolute numbers may not have decreased), allowing average farm size to increase despite the pressures of population growth; agriculture remains dominated by small groups of households owning more than 1.0 ha of *sawah* who (although comprising less than 9% of all households in the nine sample neighborhoods) own more than half the available *sawah*.

Some summary statistics on the levels and sources of income of the different landownership categories—departing from our usual practice and aggregating the nine sample neighborhoods—are shown in Table 13.11. Space does not allow detailed discussion of the distribution of nonpaddy agricultural activities and incomes (other seasonal crops, tree crops, livestock, poultry, and aquaculture) which together make up about one-third of the 'own farm' incomes shown in the table. (These are detailed for each sample neighborhood by landownership class in Abunawan Mintoro 1984:appendix 8.) Their overall distribution reflects

TABLE 13.11. *Summary: Distribution of Farm and Nonfarm Incomes Among Households by* Sawah *Ownership Class: 1981*

	Own farm income			Agricultural wage income		Nonfarm income		Total income	
Sawah ownership class	*% of all households* *(1)*	*Average* *(2)*	*% of total* *(3)*	*Average* *(4)*	*% of total* *(5)*	*Average* *(6)*	*% of total* *(7)*	*Average* *(8)*	*% of total* *(9)*
1.0 ha and above	8.5	1222	50	5	1	916	23	2143	31
0.5–0.99 ha	10.4	289	15	12	3	410	12	711	13
0.25–0.49 ha	15.8	212	16	25	10	360	17	598	16
Less than 0.25 ha	16.2	105	8	46	18	180	9	333	9
Nonowners									
Engaging in agricultural wage labor	41.6	50	10	67	68	255	31	372	26
Not engaging in agricultural wage labor	7.5	0	0	0	0	405	9	405	5
All households	100.0	205	100	41	100	382	100	589	100

Source: Resurveys, 1981 (Abunawan Mintoro 1984:app. 8).
Note: Aggregate for the nine sample neighborhoods, n = 1113 households; incomes in Rp thousands per year.

that of paddy incomes, resulting in the highly unequal pattern of agricultural income distribution shown in Table 13.11: 8.5% of households between them command 50% of all farm incomes. In the sample villages a 'poverty line' income (if defined as many authors have done as per-capita incomes below 320 kg milled-rice equivalent per year) in 1981 amounted to approximately Rp. 62,000 per capita or Rp. 280,000 per household at prevailing local rice prices. As may be seen in Table 13.11, the larger landowning households with more than 1.0 ha of *sawah* are the only group whose *agricultural* incomes far surpass this level, providing a substantial surplus potentially available for further expansion in agriculture or other activities. On the other hand, while agriculture now provides average incomes (in own-farm and wage income) only slightly below 'poverty line' levels in the sample neighborhoods as a whole, the 65% of households without land or with holdings less than 0.25 ha command agricultural incomes so far below this level that involvement in nonfarm activities is necessary as a matter of survival.

With the decline of share tenancy and the growing tendency for land-lease transactions to occur between landowning households, we can discern more clearly the emergence of an elite group of prosperous larger farmers or more accurately farm managers, sometimes supplementing their own holdings with land lease and mortgage, who control a large part of land and farm incomes and also provide the bulk of wage employment in what is primarily a wage-labor-based production system. (As we have seen in Table 13.8, more than 80% of all preharvest labor inputs in paddy production are made by hired labor in most villages, and the proportion is of course still higher for the larger farms.) Although there is some 'horizontal' wage circulation between smaller farm households, the bulk of wage transactions are 'vertical' with 68% of all agricultural wages earned by members of landless households (cf. Table 13.11). The changes we have described, which represent the acceleration and crystallization of trends already long in motion rather than any abrupt break with past patterns of agrarian relations, might then allow us to speak with greater confidence of the emergence of opposing 'commercial farmer/employer' and 'landless agricultural laborer' classes (with the exception of 'backward,' rain-fed Sentul where the transition from exchange to wage labor is only just beginning), at either end of a still large but relatively declining mass of petty commodity producing small-farm households still retaining some access to land and (particularly at the lower end of the scale) supplementing inadequate own-farm incomes with wage labor (cf. Table 13.11).

Such a view may be helpful in characterizing recent changes in production relations *in agriculture*, but it is inadequate and perhaps highly misleading as a characterization of *rural* classes or class relations or the 'agrarian structure' as a whole. As we can see from the right-hand side of Table 13.11, nonfarm incomes provided almost two-thirds of all incomes in our nine sample neighborhoods taken together—in what was a relatively successful agricultural year for all villages except Wanarata—and more than half the incomes of each landownership group with the exception of those owning more than 1.0 ha of *sawah* (although the latter still command the highest absolute levels of nonfarm income). As we have seen already in Table 13.1, only Mariuk can still be considered a relatively 'purely' agricultural village, with two-thirds of all income deriving from agriculture.

To examine first the income sources of 'landless agricultural laborer' households we have eliminated a small number of landless households (15% of landless households, 7.5% of all households) whose members are not engaged in any kind of agricultural activity. These include, for example, 'young' households who have not yet inherited land from relatively wealthy parents and in-migrant households with relatively good incomes from trade or industrial or salaried employment, whose incomes are on the whole higher than those of small-farm and landless agricultural worker households (cf. Table 13.11). Thus, Table 13.12 covers only those landless households whose members obtained some agricultural wage income during the year. For these households, agricultural wages provide only a small proportion of total incomes in all cases, and nonfarm sources easily outweigh agricultural incomes with the single exception of Mariuk. Petty trade and nonagricultural (casual) wage labor (columns 6 and 8 of Table 13.12) provide significant sources of income in all villages and in some cases secure salaried jobs (Kebanggan, Sukosari, Janti) such as school attendants or in local factories. Household industries are surprisingly unimportant, and the low figures provide some support for the impression that many traditional crafts are declining under the impact of competition with urban-produced substitutes. 'Service' sector incomes are prominent only in Wanarata, and in this case mainly derived from *becak* (pedicab) driving. In only a few cases (brick and rooftile industries in Sentul, *kerupuk* [shrimp-cracker] production in Janti, metalworking industries in Geneng) does wage em-

TABLE 13.12. *Income Sources of 'Landless Agricultural Labor' Households (Percentage of Total Annual Household Income): 1981*

Village	Agriculture					Nonfarm sources							
	% of all house-holds (1)	*Own crops* (2)	*Livestock, poultry, fisheries* (3)	*Wage labor* (4)	*Subtotal agricultural income* (5)	*Trade* (6)	*Household industry* (7)	*Wage labor* (8)	*Salaries* (9)	*Services* (10)	*Other* (11)	*Subtotal nonfarm income* (12)	Average household income (Rp '000) (13)
West Java													
Sentul	26	13	2	2	(17)	17	2	48	6	8	4	(83)	193
Mariuk	64	9	13	46	(68)	6	0[b]	10	3	6	7	(32)	273
Jatisari	27	8	17	15	(40)	22	1	23	—	1	13	(60)	234
Central Java													
Rowosari	50	3	31[a]	10	(44)	13	0[b]	23	5	2	14	(56)	353
Kebanggan	49	5	3	16	(24)	19	2	33	17	1	4	(76)	300
Wanarata	25	3	2	36	(41)	9	5	19	—	14	12	(59)	316
East Java													
Sukosari	25	2	8	29	(39)	11	9	13	10	6	12	(61)	295
Geneng	54	13	1	24	(38)	18	7	17	6	2	12	(62)	365
Janti	52	4	2	5	(11)	32	1	9	35	5	7	(89)	771

Source: Resurveys, 1981 (Abunawan Mintoro 1984:app. 8).
Note: Totals may not add because of rounding.
[a] 24% fishing.
[b] 0 = less than 0.5%.

ployment in the nonfarm sector involve relations with local employers who are also major employers of agricultural labor. Furthermore, as more detailed analyses have shown (Soentoro et al. 1982:ch. 5; Soentoro 1984), large and probably increasing proportions of nonfarm incomes are earned outside the village through seasonal or continuous ('circulating' and in some cases daily 'commuting') out-migration of household members to urban centers—in petty trade, casual labor in the (then) booming construction sector, as *becak* drivers, domestic servants, etc.

Given the involvement of male and female members of 'landless agricultural labor' households in such a wide variety of activities and labor 'statuses' besides farm labor, in petty commodity production, small trade, service sector and wage work, both inside and outside the village, the landless cannot easily be categorized as a landless worker class; we could more usefully underline their semiproletarian status, with all the complex and ambiguous implications for class relations, class consciousness, and class action which that status involves. The same can also be said of the smaller farm households who supplement inadequate own-farm incomes both with agricultural wages and with a similar variety of nonfarm activities both inside and outside the village. We suppose this mobility and diversification of labor will further develop among such households as landlessness and land concentration increase, as the seasonality of agricultural wage-labor demand sharpens, and as agricultural mechanization proceeds, even if real agricultural wages remain at their new higher level for those with access to them. It is interesting to note that only among the landless and very small landowning groups (with less than 0.25 ha of *sawah*) do we find a generally negative relationship between agricultural and nonfarm incomes at the household level (Abunawan Mintoro 1984:268–270), indicating that for these groups it is the inadequacy of agricultural-sector incomes which propels their members into nonfarm activities as a survival strategy.

For the middle and larger landowning groups, on the other hand, agricultural and nonfarm incomes are positively associated (ibid.)—suggesting that in this case we are dealing with a more dynamic strategy of accumulation, in which surpluses derived from one activity are used to gain access to (and higher incomes in) the other. In Table 13.13 we provide some details for each village on the agricultural and nonfarm income sources of the small numbers of households owning more than 1.0 ha of *sawah*, who (as already seen in Table 13.11) have been the major beneficiaries of state-subsidized agricultural productivity growth

TABLE 13.13. *Income Sources of Households Owning More Than 1.0 Hectare of* Sawah *(Percentage of Total Annual Household Income): 1981*

Village	Agriculture						Nonfarm sources							
	Own crops	*Live-stock, poultry, fisheries*	*Land rent*	*Sugar-cane 'rent'*	*Wage labor*	*Subtotal agricul-tural income*	*Trade*	*Small industry*	*Salaries*	*Wage labor*	*Services*	*Other*	*Subtotal nonfarm income*	Average household income (Rp '000)
West Java														
Sentul	49	0[a]	1	—	—	(51)	—	44	—	0[a]	3	2	(49)	611
Mariuk	61	12	19	—	—	(92)	4	—	3	—	1	—	(8)	1381
Jatisari	49	1	2	—	—	(52)	22	1	2	—	5	18	(48)	1544
Central Java														
Rowosari	33	0[a]	31	—	—	(64)	11	25	—	—	—	—	(36)	1376
Kebanggan	62	3	1	—	—	(65)	2	—	21	—	—	12	(35)	2778
Wanarata	17	2	16	—	—	(35)	2	—	11	5	40[c]	7	(65)	983
East Java														
Sukosari	81	5	2	—	—	(88)	—	5	4	—	3	0[a]	(12)	4134
Geneng	25	4	13	38[b]	—	(80)	0[a]	8	2	—	4	7	(20)	2034
Janti	17	4	4	34[b]	2	(60)	—	23	10	—	—	7	(40)	2438

Source: Resurveys, 1981.
Note: Totals may not add because of rounding.
[a] 0 = less than 0.5%.
[b] Profit on sugarcane (TRI) land (see text).
[c] Vehicle hire 35%.

by virtue of their control of more than half the land. Any reader who has tried to gather information on the incomes of rural elites through surveys will understand that many income sources are likely to be underestimated, and the small sample sizes (between 6 and 10 in each sample neighborhood) may introduce further distortion; nevertheless this table provides a rough idea of the relative importance of different income sources for members of these prosperous 'commercial farmer' households. Land rent is of importance in only a few villages (Rowosari, Mariuk, and Wanarata). In Geneng and Janti, reorganization of sugarcane cultivation, while intended to replace land rental (to the sugar factories) with 'smallholder cultivation' (by the landowners), has in practice resulted in a new form of rent; land assigned to sugarcane is made over to landowner groups whose leaders organize its cultivation using wage labor, and the net income from sales is divided among the landowners who may have had no direct role in its cultivation. Only in three villages are large amounts of income derived from rural small industries employing wage labor: brick and rooftile enterprises in Sentul, *kerupuk* manufacture in Janti, and commercial rice mills in Rowosari, with a smaller contribution from blacksmithing industries in Geneng which produce small agricultural tools, knives, and vehicle springs. Trade figures prominently only in Jatisari (mainly vegetables and paddy) and Rowosari (paddy and fish); paddy traders in these villages in 1981 specialized in low-price purchasing from farmers unable to reach the minimum moisture content stipulated for guaranteed-price purchase by the cooperatives, to be resold to the cooperatives without further processing through close relations with cooperative officials willing to ignore the regulations.

In Kebanggan, Wanarata, and Janti many members of the large landowning households have secure salaried positions as schoolteachers. In Wanarata, a large proportion of their nonfarm incomes derive from the hire of vehicles. We should also mention—although the limitations of questionnaire interviews mean that it does not figure in our quantitative data on incomes—that many large landowners also derive income from usury, a major source of loans for the large proportions of households in all villages (especially in the small-farm and landless groups) who reported informal-sector debts; about three-quarters of informal-sector debts were used for income generation (to finance trade or agricultural or nonfarm production) rather than consumption (cf. Colter 1984).

In summary, the commercial surpluses deriving from large landholdings and intensified production seem to be invested in a variety of nonfarm activities, few of which are new (vehicles and rice mills are the main exceptions), although many of the 'traditional' ones are modernizing and expanding (for example, through investment in machine-pressing of rooftiles in Sentul). A similar impression emerges from examining the new assets recently acquired by large landowners; in 1981, only in three villages was any new land acquired by this group, the main other form of asset acquisition being vehicles. The large landowners, besides their substantial farm surpluses, also have the greatest access to subsidized bank credits; comprising only 3% of all households and 20% of all formal-sector borrowers, they receive more than three-quarters of all subsidized credits (cf. Colter 1984). We may suppose, therefore, that they will continue to be the dominant rural investors and follow a diversified pattern of investment in small industry, trade, agro-processing, land acquisition, and usury, balancing these against the demands of children's education and conspicuous consumption (televisions and videos, house improvement, etc.), which are the most visible sign of growing differences in wealth and life-style between rich and poor—even if the incidence of absolute poverty is itself decreasing—in a time of relatively rapid agricultural and nonagricultural income growth.

These patterns help us to understand why land concentration is not proceeding faster, even though the surpluses available to finance further concentration of landed property are increasing. On the one hand, wealthy households have many other avenues for profitable investment, and many demands for nonproductive expenditure, which compete with the alternative of land acquisition. On the other hand, the many smaller owners whose agricultural incomes do not provide reproduction at minimal levels (cf. Table 13.11) are able by participating in a variety of low-return nonfarm activities both inside and outside the village to achieve subsistence incomes without the distress sale of their 'sublivelihood' plots. These patterns, which are certainly not unique to Java, call for interpretations of agrarian differentiation processes under conditions of commoditization and productivity growth which place the phenomenon of 'part-time' farming and farm labor at all levels of the agrarian structure in more central focus.

Notes

1. Limitations of space prevent us from discussing these problems in comparability, which are described in detail in a longer version of this paper, "Agrarian Changes in Nine Javanese Villages, 1971–1981," available from the authors.

The baseline data for '1971' were obtained from two sources. The first comprised sample surveys of thirty farm households in each village from Round 5 of the Agro Economic Survey's 'Rice Intensification Study,' covering wet season 1970–1971 and dry season 1971. These samples are somewhat upwardly biased with respect to farm size due to the purposive inclusion of five 'large farmers' in the sample. The second source comprised a partial agricultural census conducted at the end of dry season 1971 in the same villages, which covered all households in two or more neighborhoods to a total of about 200 households and selected those neighborhoods in which the greatest number of the thirty sample farmers were located. These two sources are referred to as 'Sample Survey' and 'Agricultural Census' respectively in the tables. The 1981 resurveys (covering wet season 1980–1981 and dry season 1981) covered all households in groups of contiguous neighborhoods up to a total of about 125 households in each village. It is important to bear in mind that while we speak of the nine 'villages' in the text and tables we are in fact analyzing *parts* of the villages (neighborhoods), with considerable overlap but not complete correspondence in coverage between 1971 and 1981.

Since there is sometimes considerable interneighborhood variation within the rather large administrative units called 'villages' (*desa* or *kelurahan*) in Indonesia, the lack of complete correspondence between the neighborhoods surveyed in 1971 and 1981 can result in problems of comparability. We think, however, that the 1981 resurveys, trying to cover the largest possible proportion of the baseline neighborhoods, create less problems of comparison than Hayami and Kikuchi's previous (1979) resurvey in Mariuk (Hayami and Kikuchi 1981:ch. 9, in which Mariuk is referred to as 'North Subang village'); the 1979 resurvey appears to have been carried out in a quite different corner of Mariuk than the baseline surveys. Hence it is not suprising that some of Hayami and Kikuchi's conclusions about change in Mariuk differ from ours.

In addition to published monographs on most of the nine villages (cited in the text), many of the data have been incorporated in a lengthy report covering these nine villages plus an additional three villages from Java (outside the 1971 sample) and three from South Sulawesi (Faisal Kasryno 1984). In many parts of our own analysis we have returned to the original data, which results in some small discrepancies between our results and those of previous reports due to different procedures in dealing with missing data, extreme or improbable values, etc.

References

Abrar S. Yusuf, Anwar Hafid, Aris Kristianto, Nazifah Umar, and M. Soleh (1980). *Aspek Ekonomi Penguasaan Tanah dan Hubungan Agraris: Kasus*

Satu Desa di Kabupaten Subang, Jawa Barat. Bogor: Agro Economic Survey, Rural Dynamics Study.

Abunawan Mintoro (1984). "Distribusi Pendapatan." In Faisal Kasryno, ed., *Prospek Pembangunan Ekonomi Pedesaan Indonesia*. Jakarta: Obor Foundation.

Agro Economic Survey (1972). *Agricultural Census in 33 Villages Located in the Major Rice Producing Areas of Indonesia*. Research Notes 6. Bogor: Agro Economic Survey.

Collier, W. L., Soentoro, Gunawan Wiradi, and Makali (1974). "Agricultural Technology and Institutional Change in Java," *Food Research Institute Studies* 13, no. 2:170–194.

Colter, J. M. (1984). "Masalah Perkreditan dalam Pembangunan Pertanian." In Faisal Kasryno, ed., *Prospek Pembangunan Ekonomi Pedesaan Indonesia*. Jakarta, Obor Foundation.

Djoko Kustiono, Sumardi, Waluyo, Arifuddin Sahidu, and Sri Mulyono (1982). *Pola Penguasaan Tanah, Hubungan Kerja Pertanian dan Distribusi Pendapatan di Pedesaan Jawa: Kasus Desa Wanarata, Jawa Tengah*. Bogor: Agro Economic Survey, Rural Dynamics Study.

Faisal Kasryno (1984). "Perkembangan Penyerapan Tenaga Kerja Pertanian dan Tingkat Upah." In Faisal Kasryno, ed., *Prospek Pembangunan Ekonomi Pedesaan Indonesia*. Jakarta: Obor Foundation.

Faisal Kasryno, ed. (1984). *Prospek Pembangunan Ekonomi Pedesaan Indonesia*. Jakarta: Obor Foundation.

Gunawan Wiradi and Makali (1984). "Penguasaan Tanah dan Kelembagaan." In Faisal Kasryno, ed., *Prospek Pembangunan Ekonomi Pedesaan Indonesia*. Jakarta: Obor Foundation.

Hart, G. (1986). "Exclusionary Labor Arrangements: Interpreting Evidence on Employment Trends in Rural Java." *Journal of Development Studies* 22, no. 4:681–696.

Hayami, Y., and M. Kikuchi (1981). *Asian Village Economy at the Crossroads*. Tokyo: Tokyo University Press.

Masjidin Siregar and Aladin Nasution (1984). "Perkembangan Teknologi dan Mekanisasi di Jawa." In Faisal Kasryno, ed., *Prospek Pembangunan Ekonomi Pedesaan Indonesia*. Jakarta: Obor Foundation.

Mazumdar, D., and M. Husein Sawit (1986). "Trends in Rural Wages, West Java, 1977–1983." *Bulletin of Indonesian Economic Studies* 22, no. 3:93–105.

Papanek, G. (1985). "Agricultural Income Distribution and Employment in the 1970s." *Bulletin of Indonesian Economic Studies* 2, no. 2:24–50.

Retno Setyowati, Sunarsih, Isbandi, and Friston Siregar (1982). *Pola Penguasaan Tanah, Hubungan Kerja Pertanian dan Distribusi Pendapatan di Pedesaan*

Jawa: Kasus Desa Kebanggan, Jawa Tengah. Bogor: Agro Economic Survey, Rural Dynamics Study.

Sajogyo and W. L. Collier (1972). *Adoption of High Yielding Varieties by Java's Farmers*. Research Notes 7. Bogor: Agro Economic Survey.

Soentoro (1984). "Penyerapan Tenaga Kerja Luar Sektor Pertanian di Pedesaan." In Faisal Kasryno, ed., *Prospek Pembangunan Ekonomi Pedesaan Indonesia*. Jakarta: Obor Foundation.

Soentoro, Faisal Kasryno, A. Rozany Nurmanaf, Rudolf S. Sinaga, and Saiful Bachri (1982). *Perkembangan Kesempatan Kerja dan Hubungan Kerja Pedesaan: Studi Kasus di Empat Desa di Jawa Barat*. Bogor: Agro Economic Survey Foundation, Rural Dynamics Study.

Suseno, S. H., Suprapto, Husein Jamani, and Hardi Suratman (1982). *Pola Penguasaan Tanah, Hubungan Kerja Pertanian dan Distribusi Pendapatan di Pedesaan Jawa: Kasus Desa Rowosari, Jawa Tengah*. Bogor: Agro Economic Survey Foundation, Rural Dynamics Study.

Umar Wahyu Widodo, Maryunani, Sugeng Raharto, Sjaiful Bahri, and Santoso Basunarto (1982). *Pola Penguasaan Tanah, Hubungan Kerja Pertanian dan Distribusi Pendapatan di Pedesaan Jawa: Kasus Desa Sukosari, Jawa Timur*. Bogor: Agro Economic Survey Foundation, Rural Dynamics Study.

White, B., and Endang L. Hastuti (1980). "Different and Unequal: Male and Female Influence in Household and Community Affairs in Two West Javanese Villages." Working Paper 6. Bogor: Agro Economic Survey, Rural Dynamics Study.

Chapter Fourteen

Cycles of Commercialization and Accumulation in a Central Javanese Village

FRANS HÜSKEN

As noted in Chapter 12, social differentiation in rural Java and the growth of a commercial type of agriculture were neither abrupt developments brought about by the modern agricultural technologies of the 1970's nor linear progressive tendencies in which capitalism penetrated steadily into the village economy since the early years of colonial exploitation. Instead, commercialization and 'decommercialization' in the Javanese countryside seem to have been cyclically alternating responses to changing conditions of an outside market which has determined their course and pace. When the 'New Order' regime came to power in 1966 and launched its *Bimas* and *Inmas* programs for agricultural modernization, Java's rural economy experienced a new commercial impetus.

Does this imply that commercialization of agriculture had now taken up again where it left off at the outbreak of the economic depression in 1930 and that rural Java is on its way to a form of capitalist development? In order to answer these questions I will briefly sketch the development of a commercial economy and its effects on forms of agricultural exploitation and surplus extraction in a predominantly rice-growing village on the north coast of Central Java. In the past the region's development has been studied at two crucial periods—by Burger in 1928 and by Bachtiar Rifai in 1956 respectively—and the village belongs to those very few where local archives cover a time span of more than fifty years. Accordingly we are able to examine its economic history in some depth.

Tayu: A Coastal District in North Central Java

For a long time this district, which encompasses the northern part of Pati Regency (northeast of Central Java's capital, Semarang), was relatively isolated. Around 1800 much of the district was still covered with forests and it was only in the nineteenth century that the area became more densely populated and that large tracts of these forests were converted to agriculture. Because of its favorable situation on the eastern slopes of the Gunung Muria, a dead volcano, the peasants of Tayu had enough water to irrigate their rice fields the whole year round.

The fertile soils of the coastal plain made the area eminently suited to the compulsory government-administered cultivation of sugarcane in the years of the Cultivation System. Around 1850 there were two sugar mills, and approximately 425 hectares (or about 13 percent) of the best *sawah* fields were annually planted with cane. More than 2000 households from sixty villages had to work these fields through a system of corvée labor. In the more elevated areas 400,000 coffee trees had to be maintained by the inhabitants of another thirty-four mountain villages. Although cane and coffee were grown for commercial purposes, the peasantry was only indirectly linked to the outside market: both crops had to be handed over as a form of tax in kind to the colonial government which took care of the marketing and the revenues of the produce.

In the latter part of the nineteenth century, however, commercial agriculture spread rapidly in the area. After the abolition of the Cultivation System in 1870 the area planted with sugarcane doubled: the sugar mills started to rent land from the villages, but also quite a few villagers began to grow their own cane which they sold either to the mills or traded directly to the local market. In nearly the same period, Tayu became a center for the cultivation of new smallholder cash crops: peanuts (*kacang tanah*) and especially kapok, which was easily grown along the village roads and which proved to be extremely lucrative. Agriculture gradually lost much of its subsistence character. More people became involved in trade, transport, and communication, and the village economy became so substantially monetarized that in the late 1920's

Burger estimated that the villagers were receiving some 75 percent of their income in cash (Burger 1930:4).

With the introduction of commercial agriculture in the region, Chinese influence grew too. Not only was there a sixfold increase in the total number of Chinese in the district (from a mere 174 in 1867 to 1003 in 1930), but more importantly Chinese control over commercial agriculture and intermediate trade expanded considerably. In 1906 the sugar factory of Pakis passed into the hands of the Oei Tiong Ham Corporation in Semarang—the first Indonesian multinational and probably the largest economic enterprise in Indonesia's colonial history—which was involved in many different branches of business and trade in and beyond Indonesia. The Chinese also dominated the trade in rice, peanuts, and kapok. Several rice mills and all ten kapok warehouses in Tayu were owned by Chinese middlemen, who sold their produce to the Kian Gwan firm, which was part of the same Oei Tiong Ham Corporation (Burger 1930; Tjoa 1963). To get a firmer grip on production, the kapok warehouses and the sugar factory in close cooperation became active dispensers of credit by paying advances on kapok deliveries or on future land leases or on both.[1]

The economic growth that had begun after the First World War led to the establishment of several new trades in Tayu. In the northern part of the district a government rubber estate was established, while near the sugar mill in Pakis a small Dutch cement factory started operating. In the twenties Tayu could therefore be considered part of a fairly well developed commercial and monetarized economy and a relatively prosperous region that attracted many immigrants who came to work in the sugarcane fields or in kapok and rice harvesting.[2]

Although communal land tenure still existed (one-third of all *sawah* was held communally), landlessness was already a common phenomenon: 40 to 50 percent of the villagers owned no land, whereas several village heads owned up to 50 hectares and sometimes more (Burger 1930:8–10). In these conditions sharecropping and tenancy had become widespread: about one-quarter of the *sawah* was cultivated by tenants and another quarter was rented out either to villagers or to the sugar mill.[3]

However, the world economic crisis of the thirties put an end to this period of relative prosperity. The sugar factory stopped its operations—after 1932 no more land was rented and in 1933 all regular workers at the mill were dismissed. Prices of kapok slumped on the world market, and this brought a sudden cut in income for many villagers who had become dependent upon the sale of this crop. Other income sources dried up as well, as a result of the sharp drop in purchasing power: the local industries got into difficulties as sales went down and they experienced an even worse setback when government restrictions on imports led to an increase in the price of raw materials. The crisis resulted in a rapid decline of commercial agriculture in the villages and in a collapse of the land market. While it had earlier been advantageous to rent or buy *sawah*, such land transactions stopped almost completely after 1930.

The subsequent decades brought war, famine, and revolution to Indonesia, but when finally political independence was achieved, economic conditions within the area did not improve much: sugarcane had lost its importance as a major crop, though the mill resumed operations in 1951. With the exception of peanuts, the cultivation of cash crops occupied only a minor fraction of village lands in Tayu.

It was only by the late sixties that conditions changed greatly in the district. The *Bimas* program was introduced by the government and after some initial problems got under way: in 1975 more than 80 percent of all rice planted in the area belonged to one of the modern varieties. Prices for secondary crops like peanuts, cloves, and citrus went up, too, and as a result of a new government program for smallholder cultivation of sugarcane the 'classic' cash crop began to occupy more and more village land.

Gondosari: A Rice-growing Village

Gondosari is only a few miles away from the subdistrict town of Tayu. In many respects its agrarian history followed the general pattern of the region. Rice cultivation, which covers about half of its 325 hectares, is the main agricultural activity, though cash crops like peanuts, cassava, maize, kapok, cloves, and citrus play an increasingly important role in the local economy.

The distribution of land is highly uneven, as Table 14.1 shows. Only one-fourth of the villagers own a piece of land sufficiently large (more than 0.25 hectare) to live on. Three-quarters of the population more or less depend for their livelihood on working the land for others as sharecroppers (who receive one-third of the yield) or as day laborers (women getting on average Rp. 100–Rp. 150 per day, men Rp. 200–Rp. 300) or have to look for work outside agriculture.[4]

TABLE 14.1. *Distribution of Landownership in Gondosari: 1976–1977*

Socioeconomic class	No. of households	Total area owned (ha)	*Sawah* area owned (ha)
Large landowners (> 2.5 ha *sawah*)	31 (5%)	206 (56%)	161 (64%)
Rich and middle peasants (0.5–2.5 ha *sawah*)	86 (13%)	110 (30%)	71 (28%)
Small peasants (0.25–0.5 ha *sawah*)	63 (9%)	27 (7%)	13 (5%)
Marginal peasants (< 0.25 ha *sawah*)	122 (18%)	23 (6%)	7 (3%)
Landless	376 (55%)	—	—
Total	678 (100%)	366 (100%)	252 (100%)

Note: The total area includes land in other villages in the district owned by inhabitants of Gondosari and therefore exceeds the village acreage considerably.

As elsewhere in Java, it was particularly these landless and near-landless whose employment dropped greatly after the new rice technology got under way in the village in 1971. The rice intensification program put a premium on efficient farm management and consequently led to an extension of laborsaving practices. Because of the gender-specific division of labor in Javanese rice agriculture, the consequences of these 'rationalization measures' fell mainly on the women of these poorer households. In the pre–Green Revolution period a considerable part of agricultural labor was performed by women. They were generally the ones to take care of those farming operations that were most labor-intensive: the transplanting of seedlings, weeding, and harvesting. As the *Bimas* farmers tended to cut down on their expenses on labor it was particularly these operations which became obvious targets for 'rationalization.' *Tebasan* and *borongan tandur* (contracting out the harvest and the transplanting) were the most common ways of limiting the number of participants to those exclusively engaged by the contractors. Another way of cutting expenses arose in 1973, when a local entrepreneur bought the first rice-huller, thus mechanizing the highly labor-intensive process of hand pounding. Other measures and devices were increasingly brought into use, so that after one or two years all large landowners and most middle peasants had abandoned the former 'open field system' in which everyone was allowed access to the harvest.

As mentioned before, the main victims of this process were the women of the landless households, who earlier did the transplanting, harvesting, and pounding of rice and thereby made a large contribution to family income. By late 1977 their chances of employment had further diminished: two new rice-hullers had been installed (also serving neighboring villages), payment in kind for rice harvesting had dropped from a one-eighteenth to a one-twenty-second share, and the chances of earning a little extra money from peanut harvesting or petty trade were greatly reduced. In 1977 many peasants began to plant insect-resistant varieties with short stalks, and this caused a complete transformation of harvesting techniques. The harvest contractors (*penebas*) began to make widespread use of male rather than female labor. Where previously women had cut the ears one by one with a small blade (*ani-ani*), now men harvested with sickles. The rice was now threshed in the fields with home-made wooden threshing machines immediately after cutting. Where formerly, under the old harvesting systems, some 200 women could harvest one hectare a day, now an average of 25 to 30 men sufficed. In 1982, when I revisited the village, this system of threshing (*ngedos*) appeared to have become general throughout the region and nowadays the fields are nearly empty at harvest time, while only a couple of years previously this used to be the busiest period.

Opposed to this class of marginal peasants and landless there is a small but powerful group of large landowners in Gondosari. It consists of thirty-one households which own a total of 161 hectares of *sawah* and 45 hectares of dry fields (*tegal*), either in the village itself or in neighboring villages. It is this group of wealthy landowners that has benefited from the commercialization of agricultural production and the increased rice yields since the late sixties. Several of them gradually are becoming rural entrepreneurs who invest heavily in agriculture and who extend their activities to other sectors such as local rice-hullers and the trade in rice and cash crops. Moreover, it is particularly they who began limiting the number of agricultural laborers on their *sawah* and *tegal* fields and cutting back on real wages.

This control of scarce resources like land and capital has matched control of the politically important positions in the village. The locally very powerful village administration (*sarekat desa*), the members of which generally receive quite extensive salary lands while they are in office, is nearly completely in the hands of these rich landowners, as is the leadership of the local chapters of the two main political parties: the government-backed corporatist Golkar and the Muslim United Development Party.

It is on this village elite and its relations with other rural classes that I want to concentrate in the next part of this paper. First I will discuss their access to village land and labor, next their activities outside agriculture since the early seventies, and finally their mechanisms of surplus extraction.

The Village Elite and Control over Land and Labor

The existence of a small group of rich landowners who control the most central resources in the village economy is not a recent phenomenon. With a few exceptions, they all belong to the two leading factions in Gondosari, descending from nineteenth-century village headmen. These had settled in the area when the larger part of it was still covered with forest. As village chiefs, they were entitled to corvée labor from the villagers, and with it they were able to clear the forest. Thus they got possession of large tracts of valuable land, most of which was converted into *sawah* (MWO 1908).

By keeping the position of village head (and thereby access to the large acreage of salary lands that goes with it) within their families and by a policy of intermarriage, this local elite managed to maintain its position as the economic and political influentials up to the present day (Hüsken 1984). Village land registration going back to the late twenties shows that they have owned about half of Gondosari's *sawah* and approximately one-sixth of the village *tegal* fields ever since 1928 (see Table 14.2). Table 14.2 also shows that while the total number of households doubled in fifty years, the number of landless nearly tripled.[5] On the other hand the large landowners—nearly all offspring of the wealthy villagers of 1928—have been able to retain land to a large extent in their hands, despite the Javanese system of bilateral inheritance and intergenerational transfer. Not only has the village elite ensured its control over land, the most valuable form of property and the most important instrument of production in an agricultural economy, by keeping it in the family for nearly a century, but they also succeeded in extending their control, mainly by renting in land from needy villagers. In 1976 a total of seventy-eight households (12 percent of the village population) rented out 44 hectares of *sawah* and another 15 hectares of

TABLE 14.2. *Land Tenure in Gondosari (in Hectares): 1928–1976*

Socioeconomic class	1928 Households	1928 Area owned	1976 Households	1976 Area owned
Large landlords	19 (6%)	138 (63%)	31 (5%)	124 (59%)
Rich and middle peasants	61 (19%)	58 (26%)	86 (13%)	61 (29%)
Small peasants	42 (13%)	14 (6%)	63 (9%)	14 (7%)
Marginal peasants	64 (20%)	9 (4%)	122 (18%)	11 (5%)
Landless	129 (41%)	—	376 (55%)	—
Total	315 (100%)	219 (99%)	678 (100%)	210 (100%)

Note: Unlike Table 14.1, this table does not include land owned outside the village because comparable data for 1928 were not available. For convenience's sake, irrigated and nonirrigated land has been aggregated according to a general formula of 1 hectare *tegal* equaling 0.5 hectare *sawah*.

tegal fields, which means that one-sixth of the total area owned by the villagers is rented out.

It is mainly the younger members of the village elite who rent in land: sons and sons-in-laws of the wealthiest village officials, retired members of the village administration or of the local *haji*, who have received at least secondary school education. Most of them are also engaged in different kinds of trade, transport, and credit transactions or hold regular jobs as employees of the sugar factory, headmaster of the village school, or military district officer. They generally rent land cheaply from indebted peasants who afterwards continue working the lands, but now as sharecroppers.

Cash income as such is, however, not a prerequisite for commercial activities: as a result of their good connections with the district official and the administrator of the regional branch of the government bank (Bank Rakyat Indonesia), members of the elite are eligible for cheap credit which they can use for a variety of purposes.[6] Although *Bimas* credits are intended for buying the expensive new agricultural inputs, and most peasants use them that way, the money can be easily spent on other purchases like pickups, minibuses, trade, usury, or, in some cases, for renting land.

Sharecropping appears to be the main way in which landlords exploit their land and therefore also the main relation of production in the village. None of the large landowners, and only a few of the rich peasants, cultivate the land themselves or through wage laborers only. Nearly half (48 percent) of the total *sawah* area and 60 percent of the *tegal* fields are worked by sharecroppers. This proportion of land not cultivated by its owners might seem high (compare, for example, the nine villages analyzed by White and Wiradi in the preceding chapter, Table 13.6) but as early as 1928 sharecropping and renting were already the major practices in Gondosari. Burger (1930:78) mentions that in that year 44 percent of the village lands were sharecropped out, 7 percent rented to covillagers, and 16 percent rented to the sugar factory. The total of land not cultivated by its owners (67 percent) is nearly identical to the proportion of land sharecropped or rented in Gondosari in 1976 (when it reached 68 percent).

The most typical form of sharecropping in Gondosari is called *morotelu,* according to which the yield is divided in a ratio of 1:2, meaning one-third for the sharecropper and two-thirds for the landowner. Usually, the sharecropper is accountable for all costs, except for a part of the costs of seeds, fertilizer, irrigation, and harvesting: two-thirds of this is borne by the landowner, who has no further expenses apart from the relatively low land tax (*Ipeda*). Besides tilling the land, caring for the growing paddy, weeding and irrigating the *sawah,* and supervising the harvest, the sharecropper is responsible for transporting the harvested rice to the owner's house and sometimes for drying the paddy afterwards.

But sharecropping is more than a contractual relationship between two parties. In Gondosari it includes all elements of the asymmetrical and diffuse patron-client bond and of personal dependency. Typically the sharecropper and his family have to perform all sorts of jobs in and around the house and yard of the landowner without pay, and when occasion arises (for example, during national or village elections) they have to prove their political loyalty by lending support to the patron's party or candidate. On the other hand, a sharecropper can appeal for financial assistance in the form of an interest-free loan whenever he is in trouble and unable to meet production costs or when he needs money for some other purpose. Since loans in the village normally bear an interest rate of 20 percent per month and are granted only when there is adequate collateral, this element of credit, together with the guarantee of employment, is very important for the sharecropper.

Even though there are some middle peasants engaged in sharecropping (but in those cases they generally cultivate either their parents' or other

TABLE 14.3. *Landownership and Sharecropping in Gondosari: 1976*

Socioeconomic class	*N*	No. of sharecropping households	Total *sawah* area sharecropped (ha)	Total *tegal* area sharecropped (ha)
Large landowners	31	2 (6%)	2	—
Rich and middle peasants	86	20 (23%)	14	3
Small peasants	63	18 (29%)	9	3
Marginal peasants	122	60 (49%)	26	13
Landless	376	198 (53%)	70	50
Total	678	298 (44%)	121 (=48%)	69 (=61%)

relatives' *sawah*), it is mainly the marginal peasants and the landless who depend upon sharecropping as a major way of earning a living, as Table 14.3 shows.

As I noted above, several forms of wage-labor arrangements are used in Gondosari for transplanting, plowing, weeding, and harvesting (which are paid for either in cash or in kind). However, no landowner depends solely upon them for the cultivation of his crops. Small peasants work the land themselves with the help of their family and hiring additional wage laborers for specific tasks. When the land is given out to a sharecropper, it is the latter who has to pay the wages for transplanting and plowing. Only in a few cases of commercial garden cultivation of citrus and cloves—the most expensive cash crops—some landowners who have recently planted these trees rely for the major part on wage labor for weeding, maintaining, and guarding these crops.

Access to Nonagricultural Incomes

Although agriculture is by far the most important source of income for the large landowners (accounting for 80 to 85 percent of their total annual income), only a small group within the village elite confines itself to agriculture alone. This group consists of elder couples who have not yet transferred their land to their children and also comprises female-headed households with only one son or son-in-law who manages the farm. All other landlords have important nonagricultural sources of income: either as trader in agricultural products or as owner of rice-hullers, trucks, vans, and local shops. More than a quarter of them hold a regular salaried position with the sugar mill, government or Islamic schools, the military area command, or the district's transmigration service. Because of their monthly salaries, these large landowners have cash which they use either for renting in land (see above) or for engaging in trade.

Ever since the nineteenth century intermediate trade has been the domain of Chinese merchants. Their position in postindependence Javanese society, however, has changed considerably. The famous Oei Tiong Ham Corporation which had dominated the regional economy in Tayu in prewar years was taken over in the late fifties by the military staff of Central Java's Diponegoro Division. National economic problems and mismanagement led to a near standstill in sugar production. Similar problems occurred in other commercial sectors, particularly when in 1960 the Chinese were forbidden by law to engage in rural trade. Some of them, however, have kept their business networks intact and proved to be useful allies of the local elites when trade began to boom again in the early seventies. Also among the wealthy villagers of Gondosari buying and selling of crops grown in the village (rice, peanuts, cassava, chillies, and kapok) has become a popular and profitable way of making money. Generally this kind of local trade is carried on in close cooperation either with Chinese merchants in the subdistrict towns or with officials of the regional government who have discovered trade as a useful supplement to their monthly salaries. With trade, trucks and pickups have made their appearance in the village and a regular traffic has developed, though the unpaved road system in the village has not yet been adapted, resulting in occasional flat tires or broken axles for the overloaded trucks and vans.

Rice, peanuts, and cassava are now usually bought through the *tebasan* system, already described above, although landlords do not necessarily engage in this harvesting practice themselves. Instead, a favored sharecropper or overseer buys the crop, organizes the harvest, and has the crop carried to the granaries of the landlords. There the crops are collected, sun-dried, and sold to merchants in nearby towns. The most successful traders control the greater part of the peanut and cassava harvest in Gondosari and some of the surrounding villages and buy up large amounts of the locally grown rice.

While *tebasan* is the most common form of agricultural trade, it is not the only one: some landlords also behave as 'penny capitalists' by buying crops piecemeal from small peasants. One of them is even engaged in buying up the *ganyahan*—the small share (on average half a kilo) that peanut harvesters receive as payment in kind—from the laborers when they return from the fields. At harvest time he was thus able to collect one to two tons of peanuts, which he then dried and sold to the dealers in Tayu.

The owners of the local rice-hullers (including the former village headman, his brother-in-law, and a nephew of his) have the best opportunities to make money in the village rice trade because of their near-monopoly in processing and their ownership of trucks and pickups for bulk transport. They have succeeded in accumulating considerable capital funds by entering the intricate network of regional trade, which reaches as far as the provincial capital of Semarang, some hundred kilometers away.

A remarkable branch of the local trade that has also become remunerative for several rich peasants is the trade in teak wood. Gondosari adjoins the state-owned teak forest and has a long tradition of illegal wood-felling by villagers. The Forestry Service officially tries to prevent these thefts, but because it is understaffed and because it pays well to overlook the small bands of men who at night enter the forest to cut down the expensive trees, timber has become an important source of income, particularly for landless households living near the forest.[7] Most of these men work from time to time in the construction of houses or in cabinet making and carpentry. In order to get cheap raw materials they used to form small bands and fell one or two trees at a time, which they then sawed in the forest. As living conditions have become harder for them, they have gone into the forest more often and sold the wood to others. To avoid interference from the Forestry Service, every now and then they gave a cabinet or teak beam to the forester. But when a new administrator was appointed in the mid-seventies, big changes took place. The new official demanded payment in advance: if he was informed of a nocturnal raid and if he had received sufficient compensation, he was prepared to give the men a free run. If not, armed patrols were sent out and the woodcutters were arrested and sent to jail. Those who could not afford to pay this compensation either had to stop or run the risk of several months in prison. Thereupon others, who up to then had not been involved in the teak trade but who had money, entered the scene. Most of them were relatives of the rich peasants and could easily borrow to pay off the forester. They then hired a band of experienced timber thieves at Rp. 200 per person a night to cut down the trees and carry them to the village. There others sawed them on a contract basis (*borongan*).

Because of this, the landless who used to work on their own account have gradually stopped doing so and have turned to the new patrons who control the market. These have established an efficient organizational network including even the district police, who turn a blind eye at the trucks when the timber is transported to a cabinet-making factory 10 kilometers from Gondosari.

It is not only the village rich who are engaged in a diversity of income-earning activities, for if we now turn to the other end of the social spectrum in the village we find marginal and landless peasants working in a variety of jobs. Most sharecrop other people's land or work as agricultural laborers or harvesters. But all save a few must supplement their income from other sources: either by joining the 'timber trade' (at least one-eighth of all households), by collecting firewood to sell in the village or on the local market in Tayu (10 percent of the villagers), or by working as carpenters, bricklayers, or day laborers in the sugarcane fields. They also live by petty trade in rice and cassava or by preparing food and snacks for sale. What all these sources of income have in common is that they require long hours of arduous toil as returns to labor are low, varying between Rp. 10 and 25 per hour (see White 1976a).

Household Budgets: Composition and Living Standards

These differences in *sources of income* among the villagers are reflected in widely divergent *levels of income*.[8] Table 14.4 shows considerable income inequalities between better-off peasants (27 percent of all households) and the other three-quarters of the population. Per consumer unit,[9] these rich peasants have an average income that is three times greater than that of small and middle peasants, five times greater than that of sharecroppers, and more than eight times greater than that of marginal and landless peasants. If we take the general Indonesian poverty line as a crude yardstick, then marginal and landless peasants in Gondosari are slightly below subsistence level.[10] Sharecroppers are in a somewhat better position, while the other classes are all well above subsistence level and even have a large surplus.

TABLE 14.4. *Percentage of Net Income from Different Sources (Sample Households)*

Socioeconomic class	Agriculture (1)	Home garden (2)	Petty trade/ artisan (3)	Agricultural wage labor (4)	Nonagricultural wage labor (5)	Firewood/ timber collection (6)	Gifts (7)	Cattle/ poultry (8)	Salaries (9)	Intermediate trade (10)	Other (11)	Total monthly Rp. income per consumer unit (12)
Rich peasants (> 1 ha)	82	7	—	—	1	—	2	4	1	2	1	23,845
Middle/small peasants (0.25–1 ha)	50	13	10	6	2	—	7	7	2	1	2	7,880
Sharecroppers	33	10	13	17	7	2	14	2	—	1	—	4,595
Marginal peasants (< 0.25 ha)	15	6	11	39	17	3	7	2	—	—	—	2,920
Landless	—	12	—	27	23	18	19	—	—	—	—	2,915

Note: Based on a household budget survey in Gondosari in 1976–1977 ($N = 72$). Figures include income earned by all household members.

TABLE 14.5. *Average Monthly Expenditures (Rp.) per Consumer Unit*

Socioeconomic class	Food	Household necessities	School	Ceremonial	Other	Total	Balance
Rich peasants	4,700	990	1,865	1,170	1,905	10,630	+13,215
Middle/small peasants	2,975	810	825	595	485	5,690	+2,190
Sharecroppers	2,650	685	350	365	245	4,295	+300
Marginal peasants	1,965	485	160	140	255	3,005	−85
Landless	2,040	545	125	160	10	2,880	+35

Note: Household necessities include kerosene, cleaning material, cooking oil, sugar, and other processed foodstuffs. Ceremonial expenses include life-cycle rituals and ceremonial feasts (*slametan*) at the beginning of the planting season (*sajen tandur*), before harvesting (*sajen wiwit*), and the veneration of dead ancestors (*ruwahan*). Other expenses cover clothing, health, housing, and entertainment.

Expenditure patterns reveal a similar variation. Most villagers spend most of their income on food and other means of subsistence (see Table 14.5): the three lower strata spend between three-quarters and four-fifths of it on food, household essentials, and fuel. Only the rich and middle peasants have a large amount left over at the end of the month—varying from Rp. 2,000 to Rp. 13,000 per consumer unit.[11] These surpluses are even more impressive than they appear at first sight, since the richer classes generally consist of rather large households (4.4 and 3.6 consumer units respectively). Rich peasants' household surpluses can therefore be as high as Rp. 58,000; middle peasants' are on average Rp. 8,000 a month.

As stated before, incomes differ between classes in composition as well as size. Prosperous villagers derive nearly all their income from farming rice, peanuts, and other cash crops. Others rely on a much greater range of income sources with far lower returns to labor: petty trade, handicrafts, wage labor, collecting firewood, timber thefts, and, surprisingly, gifts. On the basis of their respective income sources and living standard a trichotomy of the village population can be constructed: (a) rich, middle, and small peasants, who can live rather comfortably off their own farming; (b) sharecroppers, who can make a living by working the land of others, supplementing their income by petty trade and wage labor; and (c) the landless and near-landless, who can hardly make ends meet and who depend for most of their income on wage labor, timber thefts, and charity from better-off neighbors or relatives.

Does this class division tend toward the classic dichotomy between owners of the means of production and proletarianized labor (with the sharecroppers as a kind of 'labor aristocracy')? I shall address this question in the last section of this paper. But first I look at the problem of local appropriation and exploitation and at the mechanisms of surplus extraction available to the village elite.

Where Do Surpluses Come From?

Deere and De Janvry (1979:607–608) have listed several mechanisms by which surplus may be extracted from peasants. These are through the private appropriation of land (rents in labor services, in kind, and in cash); through the market (for labor, products, and money); or through the state (taxation). As I am primarily concerned here with differentiation within the village, I will confine myself to intravillage extraction mechanisms, leaving aside the supravillage sphere mentioned by Deere and De Janvry: those via unfavorable terms of trade for agricultural produce (which in one way or another affect all villagers) and taxation by the state (as land taxes are still relatively low: between Rp. 4000 and Rp. 6000 per hectare for rice fields, and between Rp. 2000 and Rp. 4000 for *tegalan*).

Rental of land in cash is very common. After the abolition of the Cultivation System in 1870, the sugar factory became an important renter of village lands (contracting between 15 and 20 percent of the best *sawah* in the late twenties), though after independence its role diminished considerably: presently the mill rents not more than 4 to 5 percent of Gondosari's *sawah*. Also villagers have paid one another cash rents for the use of land since the early twentieth century, and this is now on the increase. While 14 hectares were thus rented in 1928, today the figure has risen to 44 hectares—i.e., about one-sixth of the entire village acreage (see above). It is mainly marginal peasants who rent out land in this way; as they do it for the most part because they need money urgently they generally have to accept fairly low rents. On the other hand, several members of the village administration who lost heavily by speculating in trade or in election campaigns also had to rent part of their salary land at cheap prices to their creditors, either other wealthy villagers or townspeople who in recent years began buying or renting land in their natal villages. Under such conditions rental of land is not a mechanism of surplus extraction by the owner from the renter-cultivator. On the contrary, renting out land functions more as a 'credit transaction' in which the owner is the weaker party and the renter is typically the person able to set the terms of the contract. Rentals of this sort lasting over several years (i.e., as collaterals for debts) are often a disguised form of usury. People in this position can lose control over their land completely when they are unable to repay their debts, though as an act of 'charity' they may be allowed to work it as sharecroppers. Since open usury is frowned on by the villagers and since it is expressly forbidden by Islamic and national law (though this does not prevent its actual practice in the village, of course), this method of gaining control over land is an attractive alternative for some local moneylenders. Under this cover, the head of the village *madrasah* (Islamic school) was able to rent more than 10 ha from twenty-six different owners who for one reason or another had to borrow money from him.

Even though rental of land provides a convenient means of appropriation, the main mechanisms for extracting surplus are rents in kind (sharecropping), wage labor, and corvée labor. As noted above, unpaid labor provided by client households on the *sawah* of their landlord, in his garden, or in his house is very common. Such clients are mainly sharecroppers: for them these labor services, in which their wives and children also have to join, are a crucial way of obtaining or retaining the right to work their master's land. There are several examples in recent history of sharecroppers unwilling to perform these unpaid services who were fired instantly, and needless to say this discourages 'disobedience' among the rest. Although the work as such may not be particularly arduous (cleaning the house, mending fences, drying rice, guarding the landlord's house and crop at night, or helping out at festive occasions), it has priority and may therefore prevent those who perform it from taking advantage of other opportunities to earn money elsewhere.

But sharecroppers are not the only villagers liable to corvée services. Corvée can also take other, less obvious forms. The better-off nearly always have one or more servants—often children of client families or of impoverished relatives—who are paid next to nothing save food and shelter. The girls (*batur wadon*) do the homework while the boys (*bujang* or *bocah*) run errands, herd cows, buffalo, and goats and are permanently available around the house. These *bujang* are also frequently used by their patrons to get their lands cultivated or their crops processed at low cost. Yet another category of 'bonded labor' are the *wong mondok* or *numpang*, who are allowed to build a shed in someone's yard in exchange for putting themselves wholly or mainly at the disposal of that household. Such people are often from outside the village and thus have no relatives nearby upon whom they can rely.

The other two mechanisms for extracting surplus in Gondosari are rents in kind (sharecropping) and wage labor. More than half of all work in the *sawah* is performed by wage laborers, and one-fifth by sharecroppers. This suggests that wage labor is the most common production relation in the village. But that is only true if we include the many female harvesters who are paid in kind. As we have seen, these women are now rapidly being replaced by a much smaller number of male laborers—cutting harvest employment by some 70 to 80 percent. So the number of people at present engaged in some form of wage labor is evidently falling.

In 1977, nearly all rice was still harvested by women, though *tebasan* practices already had reduced the number of participants as well as the wages in kind (*bawon*). It was nothing new for the female harvesters' wages to fall and even to fall sharply. In the twenties they were approximately 15 percent of the harvest, in the fifties they had fallen to 8 percent and in 1977 they were less than 5 percent. In the same period the number of participants had been restricted. Both peasants and harvesters have tried to cancel out the oversupply of labor that resulted from net population growth and impoverishment by not allowing non-villagers—chiefly those itinerant harvesters who used to come from the poorer regions in the residency (cf. Burger 1930:28)–to join the harvest. In 1977 the income a woman could earn by a day's work harvesting rice amounted to one or one-and-a-half kilo of paddy (which then had a value of Rp. 60–Rp. 90). This wage was much lower than that paid for other female tasks such as transplanting or weeding for which they generally received Rp. 200 a day. Nevertheless many women were still eager for such work, since the rice harvest used to offer a source of income for all female members of the household (including old women and girls) over many weeks at a stretch. The total daily income of a household with several women could therefore be relatively high.

The number of (mainly male) laborers who are paid in cash (with an occasional simple meal and a few cigarettes) is much smaller. Such people are hired to hoe, plow, harrow, repair dikes and irrigation canals, weed, and perform various other tasks in the fields. The work is highly sea-

TABLE 14.6. *Average Number of Labor Days and Amount of Chemical Inputs per Hectare of* Sawah

Socioeconomic class	N	Family or reciprocal labor	Wage labor, cash	Wage labor, kind	Total labor	Chemicals (Rp.)	Average gross yield (tons)
Rich peasants	3	178	43	106	327	8,567	4.81
Middle/small peasants	15	191	31	120	342	9,045	4.49
Sharecroppers	22	139	45	176	360	9,786	4.36
Marginal peasants	4	344	36	107	487	11,264	3.51
Average		166	40	151	357	9,508	4.40

sonal, and as the supply of labor is large, people rarely do it on a regular basis. Even people who see themselves primarily as wage laborers (*tani pocok* or *buruh*) can on average only find work for eight to twelve days a month, and then mainly in the peak season at the beginning of the rice cycle.[12] So not only are wages much less than the value produced, but also the unbalanced labor market causes wages to fall far below subsistence level. Since the government has not fixed a minimum wage, the amount paid to casual laborers is decided by the employer. Over time, the general wage level is adapted to inflation and the rising costs of living, but a downward trend in real wages of 10 to 20 percent is visible.[13]

Because wage labor in agriculture has been decreasing since the mid-seventies as a consequence of rationalization and extensification, sharecropping is becoming the dominant relation of production in the village applied to nearly half of the village lands. Why does this still occur on such a large scale in the post–Green Revolution agrarian economy? To answer this question we must examine the different ways in which labor and capital are combined in agriculture and consider the returns to them (see Table 14.6).

In Gondosari as elsewhere, the smaller the farm the more intensive the labor employed. More remarkably, small peasants also make greater use of such chemical inputs as fertilizers and insecticides. But intensive cultivation does not apparently lead to higher yields, since the better-off often have the best *sawah* (*ledokan*), while more than half of the marginal peasants cultivate fields that are not so well watered (*tenggeran*) and therefore try to offset the poorer quality of the soil by applying more fertilizer.

Nearly half the labor in the rice fields is performed by members of the cultivating households themselves. Only sharecroppers are an exception to this trend: not more than two-fifths of the labor on their plots is carried out by family members or by people engaged in reciprocal (*sambatan*) relations with them; for the rest, they employ wage laborers.[14] This is partly the result of the relatively small number of producers in sharecropping households who also try to supplement their income by taking opportunities to work as wage laborers or artisans.

Sharecropping arrangements are profitable both for the landlord (notwithstanding the somewhat lower yields) and the sharecropper (in spite of his low returns to labor). After deducting one-third of the yield and the sharecropper's one-third share in chemical input costs and land tax, net returns to the landlord are over Rp. 100,000 per hectare. He gets 184 days of labor in return for paying the sharecropper one-third of the yield. This amounts to approximately Rp. 300 per day, which is the same as the standard daily wage. It is, however, far less than the total wages the landlord would have to pay if he had the land cultivated by laborers under the supervision of an overseer: in such a case exploitation costs would be up to 25 percent higher.[15] Apart from cheap labor on his *sawah*, the landlord also receives other benefits (unpaid labor services and political support), the value of which is hard to estimate.

For the sharecropper, this arrangement guarantees him a low but steady round-the-year income. Average returns to labor of Rp. 258 per day are far higher than other landless peasants would get, since the latter are only incidentally able to find employment. Sharecroppers are therefore 'privileged': their income allows them to save some money or to spend it on durable consumer goods (see Table 14.5).

Although there are clear differences between sharecropping, employing wage labor, or requiring corvée services for the people subjected to them, these are all definite intravillage mechanisms for extracting surplus, generally controlled by the rich landowners. Their crucial control of land and capital as well as of trade and credit, and their well-established political position backed by central and regional government (and military) support, gives them the opportunity to accumulate these surpluses without much opposition from the poorer villagers.

Is Rural Capitalism Developing in Gondosari?

After this general overview of the economic organization of the village and of the changes village society has undergone since the programs for agricultural modernization and intensification started in the early seventies, I now want to turn to the question whether these developments can be seen as indicators of the growth of agrarian capitalism in Gondosari.

There are various ways to answer this question, depending upon what we understand by capitalism. If we follow the classic Marxist line of reasoning and focus on relations of production and in particular on concentration of landholdings and the existence of wage-labor relations as a criterion of capitalist transformation of village agriculture, we are left with a rather confused picture. Undoubtedly there has been a concentration of land tenure in Gondosari over the last hundred years. This went together with increasing landlessness. At present more than half of the population owns no land and another 20 percent owns only a tiny piece. This process, however, has not led to the formation of an undifferentiated dispossessed rural proletariat living off wage labor. Before 1978, wage labor used to be the most common production relation: both landowners and sharecroppers frequently employed agricultural laborers on their fields. But rationalization of the harvest—particularly the introduction of sickles and threshing machines—caused a considerable decrease in the number of female laborers. At present, less than one-third of the work on the *sawahs* is performed by wage laborers. This runs counter to the general tendency toward 'proletarianization' generally associated with capitalist penetration of agriculture. Instead, sharecropping has come to the fore as the main form of exploitation. It accounts for about half the village acreage and for 44 percent of its households. Although the local term for sharecropping (*mburuhaken*—to have a laborer work the land) has definite wage-labor connotations, it can hardly be called a capitalist production relation: in varying degrees, it still has all the peculiarities of 'bonded labor,' implying personal as well as household dependency on the landlord.

There are no signs that sharecropping has lost its importance over the last four or five decades as a way for landowners to have their land cultivated or that it has given way to wage-labor agreements. On the contrary, it has increased absolutely from 44 percent of the total acreage in 1928 to 52 percent in 1928. The number of landless households working as sharecroppers has also grown from 125 in 1956 to 198 in 1976 (next to the 79 small and marginal peasants who are employed in sharecropping too).

Apart from this, a new form of sharecropping has emerged alongside the old *morotelu* contract. According to this new form, which has been operated above all by the 'entrepreneurial' landowners, the sharecropper (now called *wong buntut*, the 'tail man') is expected to perform all agricultural tasks, except plowing and harrowing, while the landowner takes responsibility for all other expenses. After the harvest the *wong buntut* receives a share of one-ninth or one-twelfth of the yield, the rest going to the owner.[16] Far from disappearing, sharecropping turns out to be on the increase—although sometimes in a new and harsher form—and there seem to be no problems in incorporating this 'precapitalist relation of production' into a newly developing commercial economy.

The main reason for this is that under the given circumstances of land scarcity, abundant labor, and a climate of political repression, a sharecropping contract is still very attractive to the landless peasants. It assures them of a longer period of work and qualifies them to seek help from their landlords when times become hard. For the landowner, sharecropping is lucrative from a financial and managerial point of view as it would cost him much more to hire day laborers to cultivate his land. In addition, the sharecropping system means that he only has to pay labor costs after the rice is harvested. In case of total or partial failure, he either does not have to pay for the labor used in production or has to pay for only part of it. Then these expenses are all borne by the sharecropper, who even after a poor harvest still has to give two-thirds of the yield to his landlord. Another advantage of the sharecropping contract for the landlord is that it enables him to devote his time to other things (trade, village politics, leisure), since he can confine himself to a few general instructions to his sharecroppers and an occasional visit to the *sawah* to inspect the crops. Moreover, the families of his sharecroppers provide him with a reservoir of free labor that he can tap whenever necessary. Finally, his sharecroppers form a loyal group of followers upon which he, his relatives, and his friends can depend for support during national elections or elections for a new village chief. Since he still exercises control over agricultural production (it is he who decides what rice varieties are planted, if and how much fertilizer is to be used, and when the harvest is to take place) and since he is free to fire sharecroppers whom he thinks do not work hard enough, he can rest assured that the yield will be maximal.[17]

Although both parties see benefits for themselves in a sharecropping contract, this does not mean that the 'terms of trade' between them are not liable to change. The local history of Gondosari provides many examples of how the relative bargaining position of the sharecroppers vis-à-vis the landowner deteriorated as a result of processes such as population growth, increasing land scarcity, political mobilization, state intervention, and growing dependence on a market and thus on fluctuating market prices. At the beginning of the twentieth century, the costs of cultivating the land fell mainly on the owner, and the crop was divided fifty-fifty. Gradually, as the village population increased to its present number of over 3000, the burden of such costs shifted towards the sharecropper, so that the latter now bears nearly all of them; on the other hand, the owner has come to claim an increasing proportion of the harvest.

The introduction of the new varieties has increased the burden on the sharecropper still further. Owners oblige their sharecroppers to plant the new rice and employ associated inputs. One-third of these expenses falls on the sharecropper, and although he can seek a loan from the landowner in the form of an advance on his share of the harvest, this still has the effect of adding to his burden. The new varieties also require more attention and labor, since the seedlings have to be planted in straight rows and demand more frequent weeding. The sharecropper is thus forced to spend more time and effort (and, where day laborers are needed, more money) on activities from which two-thirds of the returns accrue not to him but to the landowner. Several landlords have recently engaged middlemen or overseers to exercise a stricter control over their sharecroppers. Some of these men came from outside the village, but in most cases a trusted villager was given this position of 'foreman.' They have to keep an eye on the various stages of rice cultivation, but most of all on the harvest, when they prevent a sharecropper from taking more than his due share. In such a situation the traditional patron-client bond has almost completely given way to a contractual relationship between the two parties. The sharecropper, aware that numerous other landless peasants are ready to step into his place, has no choice but to conform to the new demands upon him.

From the point of view of production relations we can therefore not speak of rural capitalism in Gondosari. We might, however, apply another definition of capitalism by looking at economic behavior. We could then follow Fegan's appeal to revive the concept of *rent capitalism* as coined by Bobek (1962; see Muller 1983). This type of capitalism 'arose through commercialization and the transformation, undertaken in a plain profit-seeking spirit, of the original lordly (or feudal) claims on income from the peasant and artisan under-strata' (Bobek 1962:234). It must be distinguished from modern *productive capitalism* 'in that it was not linked with production, but rather was satisfied with skimming off its proceeds' (Bobek 1962:237). Rent capitalism therefore "leaves intact the petty scale of production, the old native technology of production, and the internal organization of the productive unit. [The rent capitalist] expands operations by using money to gain claims on the product of more petty units of production via mortgages, purchases, and loans. He reserves his capital for speculation in products and land and for making loans to petty producers for materials of production, instruments of production, for luxuries, for subsistence, for ceremonies, and for emergencies. All these give the rent capitalist claims on the product of the petty producer" (Fegan 1981:2).

There are many similarities between the economic behavior of the village elite in Gondosari and the typical behavior of the rent capitalist described above. Sharecropping, rentals of land, usury, and corvée labor fit easily into this framework. An impressive part of the large surpluses of the wealthy peasants are used to buy expensive consumer goods: cars, motorbikes, television sets, generators, and other goods which can hardly be said to have productive value. On the other hand, we also saw that landowners were investing part of their capital in agricultural production and that they were pushing their sharecroppers either directly or indirectly through their overseers to increase their yields as much as possible. They were also very much prepared to extend their economic ventures by investing in trade, the processing of agricultural products, and investment in transport facilities. To call them pure rent capitalists in the Bobekian sense would therefore leave too many aspects of their behavior unexplained. A better term would probably be that of 'hesitant capitalists' who in the face of their suddenly increased surplus and well aware of its fragility shrink at a too heavy involvement in capitalist investments. They know from past experiences that economic and political conditions may change rapidly, and in such a situation complete dependency on the market is a risky business.

A third line of reasoning would look at the ways in which the village economy is tied to supra-village or even national economic structures. On those levels capitalism, in the form either of sugar mills or of other agricultural processing industries, already has a long history. Its specific relation with the village economy has often been described as an interlinking of precapitalist modes of production and rural capitalism. In this process of articulation it is evident that even after the massive capital injections into Javanese agriculture during the seventies, noncapitalist relations of production like sharecropping and debt labor will still play an important role for some time to come: "During an entire period [capitalism] must reinforce these relations of exploitation, since it is only this development which permits its own provisioning with goods coming from these modes of production, or with men driven from these modes of production and therefore compelled to sell their labor power to capitalism in order to survive" (Rey 1973:15–16).

We also have to consider the special role of the Indonesian state since 1965 when the military came to power and political parties lost their influence (see Chapter 12). Since then, the government has been looking for allies in the countryside after it banned its main political contestant, the Communist Party and its affiliated Peasants' Union. The rural elites were—and still are—ready partners in such a coalition, and they have benefited from it in many ways. The favorable economic conditions under which the 'New Order' government could operate—large foreign loans and increasing oil revenues—have put enough capital at the disposal of the government to spend a substantial part of its budget on subsidizing agricultural modernization. Indonesian agricultural policy has focused upon the local elites as the main agents of both political stability and economic growth by offering them new technologies, cheap credit, and an efficient marketing system. The present rise of a village elite in Gondosari is therefore to be attributed more to its position of *anak mas* (favorite child) of the state than to its entrepreneurial capacities as such. What will happen with the village socioeconomic structure when Indonesia's present economic problems and particularly the falling oil revenues make it impossible for the government to continue spending its money on patronizing these rural elites remains, however, to be seen.

Notes

The field research on which this paper is based and several trips in recent years were financed by research grants from the Netherlands Foundation for the Advancement of Tropical Research (WOTRO) and from the Department of

Sociology and Anthropology at the University of Amsterdam. For comments on earlier drafts of this paper I am obliged to the participants to the two SSRC workshops on Rural Differentiation in Southeast Asia, and for editorial advice I would like to thank Gregor Benton.

1. *Memorie van Overgave van den Assistent-Resident J. W. Meijer Ranneft, Pati 1926–1928* (Archief Meijer Ranneft Nr. 23; Dutch State Archives, The Hague).

2. Burger (1930) estimated that in the subdistrict 35 to 65 percent of the rice was harvested by migrant laborers.

3. In the subdistrict of Tayu these figures were 34 and 40 percent; so in the region as a whole only 26 percent of the land was worked by its owner.

4. The exchange rate for the *rupiah* in 1976–1977 was US$1.00 = Rp. 460.

5. Between 1928 and 1956 the total population of Gondosari grew from 1350 to 3138. As the number of households rose in these years from 315 to 678, this indicates that at present households are slightly bigger than fifty years ago (having on average 4.63 instead of 4.29 members).

6. From the records of the regional branch of the Bank Rakyat Indonesia it is obvious that the main recipients of *Bimas* credits were among the wealthiest peasants in Gondosari. From 1971 to 1976 the average acreage of *Bimas* participants in the village was 2.65 hectares. For the district as a whole this average was 1.9 hectares.

7. In total, eighty-six households were regularly or irregularly involved in these timber thefts, among them thirty-two landless, twenty-nine marginal peasants, and seventeen sharecroppers.

8. The data on which this section is based are calculated on the basis of a twelve-month sample survey of seventy-two households. Although in general they give a good impression of sources and levels of income of the village population, data for the rich households can only be used as an approximation because of their small number in the sample. Only a few elite households have been included, which tends to lead to an underrepresentation of their income sources. Another well-known reason why the data have to be treated with caution is the general tendency of the rich to conceal their income.

9. For a more detailed discussion of the concept of consumer units I refer the reader to Epstein (1962), Hart (1986), Satoto and Fatimah (1976), Van den Muijzenberg (1974), and White (1976b).

10. Average household size in the sample of seventy-two households was 4.72 persons; the average number of consumer units was 3.62. Given the commonly used measure for basic needs fulfillment in rural Java of 240 kg of milled-rice equivalents (MRE) per person per year (50 percent of which will cover rice needs and the other half nonrice food and nonfood items, see Sajogyo 1974), the minimum income per consumer-unit will be $4.72/3.62 \times 240$ kg MRE =

313 kg MRE. Converted into its rupiah value, this would amount to Rp. 37,500 per consumer unit per year, or Rp. 3130 per consumer unit per month. (The midyear price of 1 kg of rice in the village in 1977 was Rp. 120.)

11. As income is highly liable to seasonal fluctuations, the expression 'at the end of the month' is not to be taken literally.

12. A gross estimate of annual wage-labor employment in the village comes to sixty to eighty full-time male laborers in rice farming and forty to fifty on the *tegal* fields. These figures are, however, rather unrealistic, since in specific periods demand for labor clearly exceeds these numbers.

13. Male wages decreased from a rice equivalent of 2.7 kg in 1956 to 2.4 kg in 1978; female wages fell in the same period from 2.1 kg to 1.7 kg (see Bachtiar Rifai 1958). These figures refer to the 'standard' wages in the village; actual practice might show some variations. As wage laborers are often in urgent need of cash, a new type of labor relation is gradually developing. In Gondosari it is known as *kontrak*: a laborer who borrows money from a landowner has to pay the loan back by working a specific number of days on the land. In this way actual wages may drop to 60 or 70 percent of the standard wage rate. See also Franke (1972), who mentions similar developments in Pemalang that are known as *ijon kerja*.

14. This is particularly obvious with the number of harvesters, which is 50 percent higher on fields cultivated by sharecroppers.

15. A crude estimate of the total costs for hiring day laborers for all work on the *sawah* would bring labor costs of 1 hectare to Rp. 67,500 (210 days at Rp. 300 plus six days of plowing and harrowing at Rp. 750). To this has to be added Rp. 4000 for chemicals which are otherwise paid by the sharecropper. This brings the total of exploitation costs to Rp. 71,500, which is approximately 25 percent more than the Rp. 56,000 which are the costs for employing a sharecropper.

16. The number of *buntut* arrangements in Gondosari amounted in 1976 to 33. As such this type of labor relation is not of recent origin in Gondosari. Bachtiar Rifai (1958:62–63) mentions the *wong buntut* as an assistant to the sharecropper, generally a younger brother or cousin who was paid with one-third or one-fourth of the sharecropper's share. The major difference with the present-day form is that now the landowner hires a *wong buntut* directly as a cheap sharecropper. For earlier references to *buntut* contracts see Soekasno (1936:185).

17. As the owner is still the one to make decisions with regard to the varieties to be grown and the inputs to be used, sharecropping in Gondosari is also from a managerial point of view a very efficient form of exploitation, contrary to what neoclassical economic theory concludes. For a general discussion of sharecropping, see inter alia Bell (1977), Cheung (1969), Mangahas (1975), and Newbery (1975).

References

Bachtiar Rifai (1958). "Bentuk Milik Tanah dan Tingkat Kemakmuran. Penjelidikan Pedesaan di Daerah Pati, Djawa Tengah." Unpublished Ph.D. thesis, Universitas Indonesia, Bogor.

Bell, C. (1977). "Alternative Theories of Sharecropping: Some Tests Using Evidence from Northeast India." *Journal of Development Studies* 13, 4:317–346.

Bobek, H. (1962). "The Main Stages in Socio-Economic Evolution from a Geographical Point of View." In Philip L. Wagner and Marvin W. Mikesell, eds., *Readings in Cultural Geography*. Chicago: University of Chicago Press.

Burger, D. H. (1930). *Vergelijking van den Economischen Toestand der Districten Tajoe en Djakenan (Regentschap Pati, Afdeeling Rembang)*. Weltevreden: Kolff.

Cheung, S.N.S. (1969). *The Theory of Share Tenancy*. Chicago/London: University of Chicago Press.

Deere, C. D., and A. de Janvry (1979). "A Conceptual Framework for the Empirical Analysis of Peasants." *American Journal of Agricultural Economics* 61, 4:601–611.

Epstein, T. S. (1962). *Economic Development and Social Change in South India*. Manchester: Manchester University Press.

Fegan, B. (1981). *Rent-Capitalism in the Philippines*. The Philippines in the Third World Papers no. 25. Manila: University of the Philippines.

Franke, R. W. (1972). "The Green Revolution in a Javanese Village." Unpublished Ph.D. thesis, Harvard University.

Hart, G. (1986). *Power, Labor, and Livelihood: Processes of Change in Rural Java*. Berkeley: University of California Press.

Hüsken, F. (1984). "Kinship, Economics and Politics in a Central Javanese Village." *Masyarakat Indonesia* 11, 1:29–43.

Mangahas, M. (1975). "An Economic Theory of Tenant and Landlord Based on a Philippine Case." In Lloyd G. Reynolds, ed., *Agriculture in Development Theory*. New Haven/London: Yale University Press.

Müller, K.-P. (1983). *Unterentwicklung durch "Rentenkapitalismus"? Geschichte, Analyse und Kritik eines sozialgeographischen Begriffes und seiner Rezeption*. Urbs et Regio. Kasseler Schriften zur Geographie und Planung, no. 29. Kassel: Gesamthochschulebibliothek.

MWO (1908). "Overzicht van den opeenhooping van grondbezit in enkele handen door inpandneming, huur of koop." Appendix 7 of the *Mindere Welvaartsrapport: Economie van de Desa (Residentie Semarang)*. Batavia: Kolff.

Newbery, D. (1975). "The Choice of Rental Contract in Peasant Agriculture." In Lloyd G. Reynolds, ed., *Agriculture in Development Theory*. New Haven/London: Yale University Press.

Palmer, I. (1976). "Rural Poverty in Indonesia with Special Reference to Java." Research Working Paper. Geneva: World Employment Programme.

Rey, P.-P. (1973). *Les Alliances des Classes.* Paris: Maspéro.

Sajogyo (1974). *Usaha Perbaikan Gizi Keluarga. Hasil Survey Evaluasi Proyek U.P.G.K.–1973.* Bogor: Lembaga Penelitian Sosiologi Pedesaan.

Satoto and Siti Fatimah (1976). "Project on the Ecology of Coastal Villages in Kendal Regency, Central Java: Methodology and Primary Results." Unpublished paper, Universitas Diponegoro Semarang.

Soekasno (1936). "Grondwoeker als Gevolg van Crediettransacties in het Pemalangsche." *Volkscredietwezen* 24:173–210.

Tjoa Soe Tjong (1963). "O.T.H.C.—100 Jaar. Een Stukje Economische Geschiedenis van Indonesia." *Economisch-Statistische Berichten* (26 June, 10 July, and 17 July 1963).

Van den Muijzenberg, O. D. (1974). *Horizontale Mobiliteit in Centraal Luzon. Kenmerken en Achtergronden.* Amsterdam: Antropologisch-Sociologisch Centrum, Afdeling Zuid- en Zuidoost-Azië.

White, B. (1976a). "Population, Employment and Involution in Rural Java." *Development and Change* 7, 4:267–290.

——— (1976b). "Production and Reproduction in a Javanese Village." Unpublished Ph.D. thesis, Columbia University.

Notes on Contributors

CYNTHIA BANZON-BAUTISTA is Associate Professor of Sociology and Deputy Director for Research at the Third World Study Center, University of the Philippines. She holds a Ph.D. in sociology from the University of Wisconsin and has done research in the Philippines on the sociology of agriculture, rural women, and overseas labor migration.

BRIAN FEGAN, an Australian, has a Ph.D. in anthropology from Yale University. He has done extensive field research in the Philippines, and his publications cover the social history of a village, population increase and the emergence of the landless rural workers, peasant movements, everyday resistance to landlord exactions, rural violence, political history of a town elite, rent capitalism, landowner resistance to land reform, social consequences of land reform and technical change, farm mechanization, and postharvest food handling. He teaches courses on Peasant Societies, Economic Anthropology, Southeast Asian Societies, and Legal Anthropology at Macquarie University in Sydney, Australia.

ANAN GANJANAPAN is Lecturer in the Department of Anthropology at Chiangmai University, where he began teaching in 1976. After graduating in Political Science from Thammasat University, he received an M.A. in Southeast Asian history in 1975 and a Ph.D. in anthropology in 1984, both from Cornell University. His thesis, "The Partial Commercialization of Rice Production in Northern Thailand 1900–1981," published in Thai, brings together his interests in northern Thai historiography and extensive field research in northern Thailand.

GILLIAN HART holds a Ph.D. from Cornell University in agricultural economics. She has conducted field research in Java and is the author of *Power, Labor, and Livelihood: Processes of Change in Rural Java* (1986). Her other publications deal with various aspects of rural labor and agrarian change and include comparative analyses of Java and Bangladesh. Currently she is engaged in research into changing gender and class relations in the Muda region of Malaysia. She is affiliated with the Institute for International Development at Harvard University and teaches in the Department of Urban Studies and Planning at the Massachusetts Institute of Technology.

FRANS HÜSKEN is a Senior Lecturer in Anthropology and Southeast Asian Studies at the University of Amsterdam. He has done fieldwork in West Java for his M.A. thesis and in Central Java for his doctoral dissertation at Amsterdam University and has conducted subsequent field and archival research into the area's social and economic history. He is the author of *A Village in Java: Social Differentiation in a Peasant Community, 1850–1980* (in Dutch, 1988) and co-editor (with Jeremy Kemp) of *Cognation and Social Organization in Southeast Asia* (1988).

LIM TECK GHEE is a Professor at the Institute for Advanced Study, University of Malaya. He is the author of *Peasants and Their Agricultural Economy in Colonial Malaya, 1874–1941* (1977), as well as numerous other publications on Malaysian agrarian economic history and contemporary Malaysian economy and society. He holds a Ph.D. from the Australian National University.

MUHAMMAD IKMAL SAID has been a Lecturer in the Anthropology–Sociology Section, School of Social Sciences, Universiti Sains Malaysia, since 1979. He holds an M.A. from the State University of New York at Binghamton and is currently working on "Household Organization and the Reproduction of Large Capitalist Farms in the Muda Area" for submission as a Ph.D. thesis at the University of Malaya. His publications include *The Evolution of Large Paddy Farms in the Muda Area, Kedah* and "Household Organization of Capitalist Farms and Capitalist Development in Agriculture" in the journal *Kajian Malaysia*.

ANDREW TURTON is Senior Lecturer in Anthropology and currently Chair, Centre of South East Asian Studies at the School of Oriental and African Studies in the University of London, where he has taught since 1970. He has carried out field research in several regions of Thailand, notably in the north, since 1968. His work has focused on peasant culture and politics. He is co-editor (with M. Caldwell and J. Fast) of *Thailand: Roots of Conflict* (1978); (with Shigeharu Tanabe) *History and Peasant Consciousness in South East Asia* (1984); and (with John Taylor) *Sociology of 'Developing Societies': Southeast Asia* (1988). From 1979 to 1986 he was associate and consultant in the Popular Participation Program of the United Nations Research Institute for Social Development.

BENJAMIN WHITE, originally from Britain, completed a Ph.D. in anthropology at Columbia University, New York, in 1976. During 1972–1973 and 1975–1980 he was engaged in research in various parts of rural Java (during the latter period as a colleague of Gunawan Wiradi in the Agro Economic Survey). His publications have focused on various aspects of rural economy and society in Java; he is also coeditor of *Rural Household Studies in Asia* (1980) and editor of *Child Workers* (1982). Since 1980 he has been teaching in agricultural and rural development at the Institute of Social Studies, The Hague, where he is currently engaged in collaborative research on rural nonfarm production and employment in West Java.

GUNAWAN WIRADI is Secretary of the Agro Economic Survey Foundation in Bogor, Indonesia. Originally from Central Java, he completed his studies at Bogor Agricultural University in 1963 and received a Master's Degree in the School of Comparative Social Sciences at Universiti Sains Malaysia in 1978. He has been involved in field studies of agrarian economy and society in Java and other parts of Indonesia since the 1960s. In addition to numerous articles he is the author of *Rural Development and Rural Institutions: A Study of Institutional Changes in West Java* (1978) and coeditor of an Indonesian-language volume of studies on land tenure in Indonesia.

Index